# Solutions Guide

*for*

## Introductory Chemistry

Steven S. Zumdahl

James F. Hall
University of Lowell

D. C. Heath and Company
Lexington, Massachusetts Toronto

Cover: Charlotte Raymond/Science Source/Photo Researchers, Inc.



Published simultaneously in Canada.

Printed in the United States of America.

International Standard Book Number: 0-669-14468-1

10 9 8 7 6 5 4 3

*Preface*

This book contains detailed solutions for the even-numbered end-of-chapter problems in your textbook. It is intended to supplement and complement your textbook, not to replace it.

Think of this book as the last chapter of an Agatha Christie mystery story. Just as you wouldn't turn initially to the last few pages of a mystery to find out "whodunit," you should not look at the solutions in this book until you have tried to solve the end-of-chapter problems yourself. To learn chemistry, *you* have to push a pencil around on paper.

When you study chemistry, first go over your lecture notes and read through your textbook to see if there is any factual material or terminology you don't understand. Then work through the example problems your professor has done in class and the examples in the textbook, paying close attention to the method of solution used for each problem. Then try again to work through the *same* classroom and textbook examples yourself *without* looking at the solutions. Once you feel comfortable with the example problems, you can tackle the end-of-chapter problems.

You will notice that the end-of-chapter problems (and the solutions in this book) are divided into the same topic sections as the textbook. This should enable you to find the specific material you want to study or review. Problems within a section are arranged in no particular order of difficulty, and no "extra hard" problems are given. If you have truly understood your classroom work and your reading of the textbook, you should be able to solve *all* the end-of-chapter problems without any difficulty.

One topic that causes many students concern is the matter of significant figures, and the determination of the number of digits to which a solution to a problem should be reported. When solving problems, we typically do *not* round off intermediate answers, but only report the *final answer* to the appropriate number of significant figures. If several intermediate answers were rounded off in a long problem of many steps, there might be a loss of precision in the final answer due to truncation errors. The solutions in this book report all intermediate answers to *one more digit than appropriate for the final answer*. The final answer to each problem is then given to the correct number of significant figures based on the data provided in the problem. *Be aware, however, that the calculator you use to solve these problems may express intermediate answers to many more digits than are given in the solutions (calculators typically express results to eight decimal places)*. If your intermediate answers do not correspond exactly to what is given in this book, check to see if the ideas mentioned above apply.

Good luck in your study of chemistry!

*Table of Contents*

## Chapter One Chemistry: An Introduction

## Introduction

2. There are, of course, *many* examples of how we make use of chemistry in our everyday life, perhaps without our even realizing that we are applying chemical principles to a problem or situation. For example, a homeowner may use an oven or drain cleaner without realizing that the cleaner works by means of a simple chemical reaction. An automobile mechanic may use a battery charger to put a "quick charge" on your car's battery without realizing that the charger causes a chemical reaction to occur in the battery. For the three occupations mentioned in the question (physician, pharmacist, farmer), some suggestions of how the person makes use of chemistry are given:
   physician: understanding biochemical processes in the cell
   pharmacist: understanding drug interactions
   farmer: understanding use of fertilizers and pesticides

## 1.1 What Is Chemistry?

4. Many beginning students would define chemistry as "this complicated and difficult subject that I am being required to take." Although at an advanced level chemistry may indeed be complicated and difficult, the basic *fundamentals* of chemistry can be understood by anyone. A knowledge of the basics of chemistry is essential for even beginning study in many fields. For example, for many students, biology is their favorite science subject in school. However, a detailed knowledge of biology is virtually impossible without a corresponding knowledge of chemistry.

## 1.2 Solving Problems Using a Scientific Approach

6. In order to analyze a situation scientifically, three things must be done: (1) the situation must be *observed* and the problem involved stated clearly; (2) possible *solutions* to the problem must be formulated; (3) the possible solutions must be tested by *experiment*.

   (1) The best way to drive to school on Friday morning. You notice that traffic is heavy at around 8:30 A.M. as office workers leave their homes (observation). You consider leaving home 15 minutes earlier to avoid this congestion (hypothesis). You try leaving 15 minutes earlier, and find that the traffic is much lighter (experiment).

   (2) Having two examinations on the same day. You realize you have two exams on the same day, one in English (in which you are doing very well) and one in math (in which you are barely holding your own) (observation). You consider studying only one hour for the English test, and devoting the rest of your time to math in the hopes that you will score passing grades on both exams (hypothesis). You divide up your study time in this manner, and do well on both exams (experiment).

   (3) Stalling your car while your little brother is aboard. You've stalled out at the busiest intersection in town with your bratty little brother in the car; you know you can't leave him in the car

while you go for help because he'll get into trouble, and yet you don't want to take him with you because he'll whine the whole time (observation). You decide to put on your emergency flashers and wait where you are for a minute or two in case you just flooded the carburetor (hypothesis). You wait two minutes, and the car starts up like a charm (experiment).

8. Answer will depend on student experience.

## 1.3 The Scientific Method

10. a. quantitative - a number (measurement) is indicated explicitly
    b. qualitative - only a qualitative description is given
    c. quantitative - a numerical measurement is indicated
    d. qualitative - only a qualitative description is given
    e. quantitative - a number (measurement) is implied
    f. qualitative - a qualitative judgment is given
    g. quantitative - a numerical quantity is indicated

12. A natural law is a *summary of observed, measurable behavior* that occurs repeatedly and consistently. A theory is our attempt to *explain* such behavior.

## 1.4 Learning Chemistry

14. Most applications of chemistry are oriented toward the interpretation of observations and the solving of problems. Although memorization of some facts may *aid* in these endeavors, it is the ability to combine, relate, and synthesize information that is most important in the study of chemistry.

16. In real life situations, the problems and applications likely to be encountered are not simple textbook examples. One must be able to observe an event, hypothesize a cause, and then test this hypothesis. One must be able to carry what has been learned in class forward to new, different situations.

## Chapter Two Measurements and Calculations

### 2.1 Scientific Notation

2. The decimal point must be moved three places, giving three as the power of ten needed.

4. Because 0.0021 is less than one, the exponent will be *negative*. Because 4540 is greater than one, the exponent will be *positive*.

6. a. 3; negative

   b. 3; negative

   c. 8; positive

   d. 6; negative

   e. 2; positive

   f. 4; positive

   g. 1; positive

   h. 4; negative

8. a. The decimal point must be moved two places to the left, so the exponent is positive 2; 529 = $5.29 \times 10^{2}$

   b. The decimal point must be moved eight places to the left, so the exponent is positive 8; 240,000,000 = $2.4 \times 10^{8}$

   c. The decimal point must be moved seventeen places to the left, so the exponent is positive 17; 301,000,000,000,000,000 = $3.01 \times 10^{17}$

   d. The decimal point must be moved four places to the left, so the exponent is positive 4; 78,444 = $7.8444 \times 10^{4}$

   e. The decimal point must be moved four places to the right, so the exponent is negative 4; 0.0003442 = $3.442 \times 10^{-4}$

   f. The decimal point must be moved ten places to the right, so the exponent is negative 10; 0.000000000902 = $9.02 \times 10^{-10}$

   g. The decimal point must be moved two places to the right, so the exponent is negative 2; 0.043 = $4.3 \times 10^{-2}$

   h. The decimal point must be moved two places to the right, so the exponent is negative 2; 0.0821 = $8.21 \times 10^{-2}$

10. a. The decimal point must be moved five places to the left; $2.98 \times 10^{-5}$ = 0.0000298

    b. The decimal point must be moved nine places to the right; $4.358 \times 10^{9}$ = 4,358,000,000

c. The decimal point must be moved six places to the left; $1.9928 \times 10^{-6} = 0.0000019928$

d. The decimal point must be moved 23 places to the right; $6.02 \times 10^{23} = 602{,}000{,}000{,}000{,}000{,}000{,}000{,}000$

e. The decimal point must be moved one place to the left; $1.01 \times 10^{-1} = 0.101$

f. The decimal point must be moved three places to the left; $7.87 \times 10^{-3} = 0.00787$

g. The decimal point must be moved seven places to the right; $9.87 \times 10^{7} = 98{,}700{,}000$

h. The decimal point must be moved two places to the right; $3.7899 \times 10^{2} = 378.99$

i. The decimal point must be moved one place to the left; $1.093 \times 10^{-1} = 0.1093$

j. The decimal point must be moved zero places; $2.9004 \times 10^{0} = 2.9004$

k. The decimal point must be moved four places to the left; $3.9 \times 10^{-4} = 0.00039$

l. The decimal point must be moved eight places to the left; $1.904 \times 10^{-8} = 0.00000001904$

12. To say that scientific notation is in *standard* form means that you have a number between 1 and 10, followed by an exponential term. The numbers given in this problem are *not* between 1 and 10 as written.

a. $102.3 \times 10^{-5} = (1.023 \times 10^{2}) \times 10^{-5} = 1.023 \times 10^{-3}$

b. $32.03 \times 10^{-3} = (3.203 \times 10^{1}) \times 10^{-3} = 3.203 \times 10^{-2}$

c. $59933 \times 10^{2} = (5.9933 \times 10^{4}) \times 10^{2} = 5.9933 \times 10^{6}$

d. $599.33 \times 10^{4} = (5.9933 \times 10^{2}) \times 10^{4} = 5.9933 \times 10^{6}$

e. $5993.3 \times 10^{3} = (5.9933 \times 10^{3}) \times 10^{3} = 5.9933 \times 10^{6}$

f. $2054 \times 10^{-1} = (2.054 \times 10^{3}) \times 10^{-1} = 2.054 \times 10^{2}$

g. $32{,}000{,}000 \times 10^{-6} = (3.2 \times 10^{7}) \times 10^{-6} = 3.2 \times 10^{1}$

h. $59.933 \times 10^{5} = (5.9933 \times 10^{1}) \times 10^{5} = 5.9933 \times 10^{6}$

14. a. $1/10^2 = 1 \times 10^{-2}$

b. $1/10^{-2} = 1 \times 10^2$

c. $55/10^3 = \dfrac{5.5 \times 10^1}{1 \times 10^2} = 5.5 \times 10^{-2}$

d. $(3.1 \times 10^6)/10^{-3} = \dfrac{3.1 \times 10^6}{1 \times 10^{-3}} = 3.1 \times 10^9$

e. $(10^6)^{1/2} = 1 \times 10^3$

f. $(10^6)(10^4)/(10^2) = \dfrac{(1 \times 10^6)(1 \times 10^4)}{(1 \times 10^2)} = 1 \times 10^8$

g. $1/0.0034 = \dfrac{1}{3.4 \times 10^{-3}} = 2.9 \times 10^2$

h. $3.453/10^{-4} = \dfrac{3.453}{1 \times 10^{-4}} = 3.453 \times 10^4$

## 2.2 Units

16. grams

18. a. mega-
b. milli-
c. nano-
d. mega-
e. centi-
f. micro-

## 2.3 Measurements of Length, Volume, and Mass

20. 100 km (see inside back cover of textbook)

22. centimeter

24. 1 kg (100 g = 0.1 kg)

26. 10 cm (1 cm = 10 mm)

28. d (1 L is slightly more than 1 qt)

30. d (the other units would give very large numbers for the distance)

32. Table 2.6 indicates that the diameter of a quarter is 2.5 cm.

$$1 \text{ m} \times \frac{100 \text{ cm}}{1 \text{ m}} \times \frac{1 \text{ quarter}}{2.5 \text{ cm}} = 40 \text{ quarters}$$

## 2.4 Uncertainty in Measurement

34. significant figures

36. The scale of the ruler shown is only marked to the nearest *tenth* of a centimeter; writing 2.850 would imply that the scale was marked to the nearest *hundredth* of a centimeter (and that the zero in the thousandths place had been estimated).

## 2.5 Significant Figures

38.
   a. one
   b. one
   c. four
   d. two
   e. infinite (definition)
   f. one

### Rounding Off Numbers

40. final

42.
   a. 0.0000324
   b. 7,210,000
   c. $2.10 \times 10^{-7}$
   d. 550,000 (better as $5.50 \times 10^{5}$ to show the first zero is significant)
   e. 200. (the decimal shows the zeroes are significant)

44.
   a. 0.5005
   b. 127
   c. 15.40
   d. 23.098

46. decimal

48. three (based on there being three significant figures in 343)

50. none (10,434 has its last significant figure in the units place)

52.
   a. 2149.6 (the answer can only be given to the first decimal place, since 149.2 is only known to the first decimal place)

   b. $5.37 \times 10^{3}$ (each of the numbers being added is known to the same level of precision). Since the power of ten is the same for each number, the calculation can be performed directly.

   c. Before performing the calculation, the numbers have to be converted so that they contain the same power of ten.
   $4.03 \times 10^{-2} - 2.044 \times 10^{-3} =$
   $4.03 \times 10^{-2} - 0.2044 \times 10^{-2} =$
   $3.83 \times 10^{-2}$ (the answer can only be given to three significant figures since $4.03 \times 10^{-2}$ is only known to three significant figures).

d. Before performing the calculation, the numbers have to be converted so that they contain the same power of ten.
$2.094 \times 10^5 - 1.073 \times 10^6 =$
$2.094 \times 10^5 - 10.73 \times 10^5 =$
$-8.64 \times 10^5$

54. a. $5.57 \times 10^7$ (the answer can only be given to three significant figures because 0.0432 is only known to three significant figures)

b. $2.38 \times 10^{-1}$ (the answer can only be given to three significant figures because 0.00932 is only known to three significant figures)

c. 4.72 (the answer can only be given to three significant figures because 2.94 is only known to three significant figures)

d. $8.08 \times 10^8$ (the answer can only be given to three significant figures because 0.000934 is only known to three significant figures)

56. a. $(2.9932 \times 10^4)[2.4443 \times 10^2 + 1.0032 \times 10^1] =$
$(2.9932 \times 10^4)[24.443 \times 10^1 + 1.0032 \times 10^1] =$
$(1.9932 \times 10^4)[25.446 \times 10^1] =$
$7.6166 \times 10^6$

b. $[2.34 \times 10^2 + 2.443 \times 10^{-1}]/(0.0323) =$
$[2.34 \times 10^2 + 0.002443 \times 10^2]/(0.0323) =$
$[2.34 \times 10^2]/(0.0323) =$
$7.25 \times 10^3$

c. $(4.38 \times 10^{-3})^2 = 1.92 \times 10^{-5}$

d. $(5.9938 \times 10^{-6})^{1/2} = 1(5.9938 \times 10^{-6}) = 2.4482 \times 10^{-3}$

## 2.6 Problem Solving and Dimensional Analysis

58. an infinite number (definition)

60. $\frac{1 \text{ L}}{1000 \text{ cm}^3}$ $\frac{1000 \text{ cm}^3}{1 \text{ L}}$

62. $\frac{1 \text{ lb}}{\$0.79}$

64. a. $8.43 \text{ cm} \times \frac{10 \text{ mm}}{1 \text{ cm}} = 84.3 \text{ mm}$

b. $2.41 \times 10^2 \text{ cm} \times \frac{100 \text{ cm}}{1 \text{ m}} = 2.41 \text{ m}$

c. $294.5 \text{ nm} \times \frac{1\text{m}}{10^9 \text{ nm}} \times \frac{100 \text{ cm}}{1 \text{ m}} = 2.945 \times 10^{-5} \text{ cm}$

d. $404.5 \text{ m} \times \frac{1 \text{ km}}{1000 \text{ m}} = 0.4045 \text{ km}$

e. $1.445 \times 10^4 \text{ m} \times \frac{1 \text{ km}}{1000 \text{ m}} = 14.45 \text{ km}$

f. $42.2 \text{ mm} \times \frac{1 \text{ cm}}{10 \text{ mm}} = 4.22 \text{ cm}$

g. $235.3 \text{ mm} \times \frac{1000 \text{ mm}}{1 \text{ m}} = 2.353 \times 10^5 \text{ mm}$

h. $903.3 \text{ nm} \times \frac{1 \text{ m}}{10^9 \text{ nm}} \times \frac{10^6 \text{ } \mu\text{m}}{1 \text{ m}} = 0.9033 \text{ } \mu\text{m}$

66. a. $908 \text{ oz} \times \frac{1 \text{ lb}}{16 \text{ oz}} \times \frac{1 \text{ kg}}{2.2046 \text{ lb}} = 25.7 \text{ kg}$

b. $12.8 \text{ L} \times \frac{1 \text{ qt}}{0.94633 \text{ L}} \times \frac{1 \text{ gal}}{4 \text{ qt}} = 3.39 \text{ gal}$

c. $125 \text{ mL} \times \frac{1 \text{ L}}{1000 \text{ mL}} \times \frac{1 \text{ qt}}{0.94633 \text{ L}} = 0.133 \text{ qt}$

d. $2.89 \text{ gal} \times \frac{4 \text{ qt}}{1 \text{ gal}} \times \frac{1 \text{ L}}{1.0567 \text{ qt}} \times \frac{1000 \text{ mL}}{1 \text{ L}} = 1.09 \times 10^4 \text{ mL}$

e. $4.48 \text{ lb} \times \frac{453.59 \text{ g}}{1 \text{ lb}} = 2.03 \times 10^3 \text{ g}$

f. $550 \text{ mL} \times \frac{1 \text{ L}}{1000 \text{ mL}} \times \frac{1.0567 \text{ qt}}{1 \text{ L}} = 0.58 \text{ qt}$

68. $9.3 \times 10^7 \text{ mi} \times \frac{1 \text{ km}}{0.62137 \text{ mi}} = 1.5 \times 10^8 \text{ km}$

$1.5 \times 10^8 \text{ km} \times \frac{1000 \text{ m}}{1 \text{ km}} \times \frac{100 \text{ cm}}{1 \text{ m}} = 1.5 \times 10^{13} \text{ cm}$

70. $1 \times 10^{-10} \text{ m} \times \frac{100 \text{ cm}}{1 \text{ m}} = 1 \times 10^{-8} \text{ cm}$

$1 \times 10^{-10} \text{ cm} \times \frac{1 \text{ in.}}{2.54 \text{ cm}} = 4 \times 10^{-9} \text{in.}$

$1 \times 10^{-10} \text{ cm} \times \frac{1 \text{ m}}{100 \text{ cm}} \times \frac{10^9 \text{ nm}}{1 \text{ m}} = 0.1 \text{ nm}$

## 2.7 Temperature Conversions

72. Celsius

74. 273 K

76. °F

78. $t_C = t_K - 273$

    a. 275 − 273 = 2°C

    b. 445 − 273 = 172°C

    c. 0 − 273 = −273°C

    d. 77 − 273 = −196°C

    e. 10,000. − 273 = 9727°C

    f. 2.3 − 273 = −271°C

80. $t_F = 1.80(t_C) + 32$

    a. 1.80(78.1) + 32 = 173 °F

    b. 1.80(40.) + 32 = 104 °F

    c. 1.80(−273) + 32 = −459 °F

    d. 1.80(32) + 32 = 90 °F

82. a. Celsius temperature = (175 − 32)/1.80 = 79.4 °C

    Kelvin temperature = 79.4 + 273 = 352 K

    b. 255 − 273 = −18 °C

    c. (−45 − 32)/1.80 = −43 °C

    d. 1.80(125) + 32 = 257 °F

## 2.8 Density

84. g/mL ($g/cm^3$)

86. volume

88. same

90. copper

92. $$\text{density} = \frac{\text{mass}}{\text{volume}}$$

a. $$d = \frac{234 \text{ g}}{2.2 \text{ cm}^3} = 110 \text{ g/cm}^3$$

b. $$m = 2.34 \text{ kg} \times \frac{1000 \text{ g}}{1 \text{ kg}} = 2340 \text{ g}$$

$$v = 2.2 \text{ m}^3 \times (\frac{100 \text{ cm}}{1 \text{ m}})^3 = 2.2 \times 10^6 \text{ cm}^3$$

$$d = \frac{2340 \text{ g}}{2.2 \times 10^6 \text{ cm}^3} = 1.1 \times 10^{-3} \text{ g/cm}^3$$

c. $$m = 1.2 \text{ lb} \times \frac{453.59 \text{ g}}{1 \text{ lb}} = 544 \text{ g}$$

$$v = 2.1 \text{ ft}^3 \times (\frac{12 \text{ in}}{1 \text{ ft}})^3 \times (\frac{2.54 \text{ cm}}{1 \text{ in}})^3 = 5.95 \times 10^4 \text{ cm}^3$$

$$d = \frac{544 \text{ g}}{5.95 \times 10^4 \text{ cm}^3} = 9.1 \times 10^{-3} \text{ g/cm}^3$$

d. $$m = 4.3 \text{ ton} \times \frac{2000 \text{ lb}}{1 \text{ ton}} \times \frac{453.59 \text{ g}}{1 \text{ lb}} = 3.90 \times 10^6 \text{ g}$$

$$v = 54.2 \text{ yd}^3 \times (\frac{1 \text{ m}}{1.0936 \text{ yd}})^3 \times (\frac{100 \text{ cm}}{1 \text{ m}})^3 = 4.14 \times 10^7 \text{ cm}^3$$

$$d = \frac{3.90 \times 10^6 \text{ g}}{4.14 \times 10^7 \text{ cm}^3} = 9.3 \times 10^{-2} \text{ g/cm}^3$$

94. $$55 \text{ mL} \times \frac{0.82 \text{ g}}{1 \text{ mL}} = 45 \text{ g}$$

96. $$m = 155 \text{ lb} \times \frac{453.59 \text{ g}}{1 \text{ lb}} = 7.031 \times 10^4 \text{ g}$$

$$v = 4.2 \text{ ft}^3 \times (\frac{12 \text{ in}}{1 \text{ ft}})^3 \times (\frac{2.54 \text{ cm}}{1 \text{ in}})^3 = 1.189 \times 10^5 \text{ cm}^3$$

$$d = \frac{7.031 \times 10^4 \text{ g}}{1.189 \times 10^5 \text{ cm}^3} = 0.59 \text{ g/cm}^3$$

98. $$5.25 \text{ g} \times \frac{1 \text{ cm}^3}{10.5 \text{ g}} = 0.500 \text{ cm}^3 = 0.500 \text{ mL}$$

$$11.2 \text{ mL} + 0.500 \text{ mL} = 11.7 \text{ mL}$$

100. a. $1.00 \times 10^3 \text{ cm}^3 \times \frac{11.34 \text{ g}}{1 \text{ cm}^3} = 1.13 \times 10^4 \text{ g}$

$1.00 \text{ m}^3 \times (\frac{100 \text{ cm}}{1 \text{ m}})^3 \times \frac{11.34 \text{ g}}{1 \text{ cm}^3} = 1.13 \times 10^7 \text{ g}$

b. $1.00 \times 10^3 \text{ cm}^3 \times \frac{2.16 \text{ g}}{1 \text{ cm}^3} = 2.16 \times 10^3 \text{ g}$

$1.00 \text{ m}^3 \times (\frac{100 \text{ cm}}{1 \text{ m}})^3 \times \frac{2.16 \text{ g}}{1 \text{ cm}^3} = 2.16 \times 10^6 \text{ g}$

c. $1.00 \times 10^3 \text{ cm}^3 \times \frac{0.880 \text{ g}}{1 \text{ cm}^3} = 8.80 \times 10^2 \text{ g}$

$1.00 \text{ m}^3 \times (\frac{100 \text{ cm}}{1 \text{ m}})^3 \times \frac{0.880 \text{ g}}{1 \text{ cm}^3} = 8.8 \times 10^5 \text{ g}$

d. $1.00 \times 10^3 \text{ cm}^3 \times \frac{7.87 \text{ g}}{1 \text{ cm}^3} = 7.87 \times 10^3 \text{ g}$

$1.00 \text{ m}^3 \times (\frac{100 \text{ cm}}{1 \text{ m}})^3 \times \frac{7.87 \text{ g}}{1 \text{ cm}^3} = 7.87 \times 10^6 \text{ g}$

## Additional Problems

102. a. $3.011 \times 10^{23} = 301{,}100{,}000{,}000{,}000{,}000{,}000{,}000$

b. $5.091 \times 10^9 = 5{,}091{,}000{,}000$

c. $7.2 \times 10^2 = 720$

d. $1.234 \times 10^5 = 123{,}400$

e. $4.32002 \times 10^{-4} = 0.000432002$

f. $3.001 \times 10^{-2} = 0.03001$

g. $2.9901 \times 10^{-7} = 0.00000029901$

h. $4.2 \times 10^{-1} = 0.42$

104. a. centimeters
b. kilometers
c. micrometers
d. millimeters

106. $36.2 \text{ blim} \times \frac{1400 \text{ kryll}}{1 \text{ blim}} = 5.07 \times 10^4 \text{ kryll}$

$170 \text{ kryll} \times \frac{1 \text{ blim}}{1400 \text{ kryll}} = 0.12 \text{ blim}$

$$72.5 \text{ kryll}^2 \times \left(\frac{1 \text{ blim}}{1400 \text{ kryll}}\right)^2 = 3.70 \times 10^{-5} \text{ blim}^2$$

108. $52 \text{ cm} \times \frac{1 \text{ in.}}{2.54 \text{ cm}} = 20 \text{ in.}$

110. $1 \text{ lb} \times \frac{1 \text{ kg}}{2.2 \text{ lb}} \times \frac{\$1}{\text{F5}} \times \frac{11.5\text{F}}{1 \text{ kg}} = \$1$

112. $°X = 1.26°C + 14$

114. $d = \frac{36.8 \text{ g}}{10.5 \text{ L}} = 3.50 \text{ g/L} \quad (3.50 \times 10^{-3} \text{ g/cm}^3)$

116. for ethanol, $100. \text{ mL} \times \frac{0.785 \text{ g}}{1 \text{ mL}} = 78.5 \text{ g}$

for benzene, $100. \text{ mL} \times \frac{0.880 \text{ g}}{1 \text{ mL}} = 88.0 \text{ g}$

total mass, $78.5 + 88.0 = 166.5 \text{ g}$

# Chapter Three Matter and Energy

## 3.1 Matter

2. states

4. container

6. far apart

8. Liquids and gases both flow freely and take on the shape of their container. The molecules in liquids are relatively close together and interact with each other, whereas the molecules in gases are far apart from each other and do not interact with each other.

10. A gas consists mostly of empty space. When a gas is compressed, at least initially, it is this empty space which is being compressed.

## 3.2 Physical and Chemical Properties and Changes

12. chemical

14. chemical

16. electrolysis

18.
   a. chemical; a chemical reaction occurs between an acid in vinegar and the dissolved protein in the milk.

   b. chemical; exposure to the oxygen of the air allows bacteria to grow which cause the chemical breakdown of components of the butter.

   c. physical; salad dressing is a physical mixture of water soluble and insoluble components, which only combine temporarily when the dressing is shaken.

   d. chemical; milk of magnesia is a *base*, which chemically reacts with and neutralizes the acid of the stomach.

   e. chemical; steel consists mostly of iron, which chemically reacts with the oxygen of the atmosphere.

   f. chemical; carbon monoxide combines chemically with the hemoglobin fraction of the blood, making it impossible for the hemoglobin to combine with oxygen.

   g. chemical; cotton consists of the carbohydrate cellulose, which is broken down chemically by acids.

   h. physical; sweat consists mostly of water, which consumes heat from the body in evaporating.

   i. chemical; although the biochemical action of aspirin is not fully understood, the process is chemical in nature.

j. physical; oil molecules are not water soluble, and are repelled by the moisture in skin.

k. chemical; the fact that one substance is converted into two other substances demonstrates that this is a chemical process.

## 3.3 Elements and Compounds

20. element

22. elements

24. different from

## 3.4 Mixtures and Pure Substances

26. a variable

28. heterogeneous

30. a. mixture

 b. pure substance

 c. mixture

 d. mixture

 e. mixture

32. a. homogeneous

 b. heterogeneous

 c. heterogeneous

 d. homogeneous

 e. heterogeneous; pond water may contain leaves, sticks, algae, small insects, and maybe a fish or two.

## 3.5 Separation of Mixtures

34. filtration

36. The solution is heated so as to vaporize (boil) the water. The water vapor is then cooled so that it condenses back to the liquid state and collected. After all the water is vaporized from the original sample, pure solid sodium chloride will remain. The process consists of physical changes.

## 3.6 Energy and Energy Changes

38. the calorie

40. As molecules of liquid water are heated, the heat energy is converted to kinetic energy: the molecules begin to move more quickly and more randomly. Eventually, as heating continues, the kinetic energy of the molecules will be so large that the attractive forces which hold the molecules together in the volume of the liquid will be overcome. The liquid will then be said to be "boiling," as individual molecules with high kinetic energies separate from the condensed liquid and enter the vapor.

42. temperature

44. Since it requires 103 J to heat the iron over a temperature change of 25° (from 25° to 50°C), it will require 206 J (twice as much heat) to heat the same sample of iron over a temperature change of 50° (from 50° to 75°).

46. a. $4.184 \text{ J} \times \frac{1 \text{ cal}}{4.184 \text{ J}} = 1.000 \text{ cal}$

b. $1520 \text{ J} \times \frac{1 \text{ cal}}{4.184 \text{ J}} = 363 \text{ cal}$

c. $8.02 \text{ J} \times \frac{1 \text{ cal}}{4.184 \text{ J}} = 1.92 \text{ cal}$

d. $23.29 \text{ J} \times \frac{1 \text{ cal}}{4.184 \text{ J}} = 5.566 \text{ cal}$

48. a. $12.30 \text{ kcal} \times \frac{1000 \text{ cal}}{1 \text{ kcal}} = 12{,}300 \text{ cal}$

b. $290.4 \text{ kcal} \times \frac{1000 \text{ cal}}{1 \text{ kcal}} = 290{,}400 \text{ cal}$

c. $940{,}000 \text{ kcal} \times \frac{1000 \text{ cal}}{1 \text{ kcal}} = 940{,}000{,}000 \text{ cal}$

d. $4201 \text{ kcal} \times \frac{1000 \text{ cal}}{1 \text{ kcal}} = 4{,}201{,}000 \text{ cal}$

50. a. $189{,}000 \text{ kJ} \times \frac{1000 \text{ J}}{1 \text{ kJ}} = 189{,}900{,}000 \text{ J}$

b. $24{,}480 \text{ kJ} \times \frac{1000 \text{ J}}{1 \text{ kJ}} = 24{,}480{,}000 \text{ J}$

c. $2.39 \text{ kJ} \times \dfrac{1000 \text{ J}}{1 \text{ kJ}} = 2{,}390 \text{ J}$

d. $19.75 \text{ kJ} \times \dfrac{1000 \text{ J}}{1 \text{ kJ}} = 19{,}750 \text{ J}$

52. Table 3.2 gives the specific heat capacity of mercury as 0.14 J/g °C.

Temperature increase = 42.0 − 37.0 = 5.0 °C

$$10.4 \text{ g} \times 0.14 \text{ J/g °C} \times 5.0 \text{ °C} \times \frac{1 \text{ cal}}{4.184 \text{ J}} = 1.7 \text{ cal}$$

54. The specific heat capacity of water is 4.184 J/g °C.

Temperature increase = 39 − 25 = 14 °C.

75 g × 4.184 J/g °C × 14 °C = 4400 J (to 2 significant figures)

56. Table 3.2 gives the specific heat capacity of aluminum as 0.89 J/g °C. 50. joules is the heat that is applied to the sample of aluminum, and must equal the product of the mass of aluminum, the specific heat capacity of the aluminum, and the temperature change undergone by the aluminum (which is what we want).

Heat = mass × specific heat capacity × temperature change

50. J = 10. g × 0.89 J/g °C × $\Delta$T

$$\Delta T = \frac{50 \text{ J}}{10 \text{ g} \times 0.89 \text{ J/g °C}} = 5.6 \text{ °C}$$

58. $$0.232 \frac{\text{cal}}{\text{g °C}} \times 4.184 \frac{\text{J}}{\text{cal}} = 0.971 \frac{\text{J}}{\text{g °C}}$$

60. Specific heat capacities are given in Table 3.2

for gold, 25.0 g × 0.13 J/g °C × 20. °C = 65 J

for mercury, 25.0 g × 0.14 J/g °C × 20. °C = 70. J

for carbon, 25.0 g × 0.71 J/g °C × 20. °C = 360 J

62. Heat = mass × specific heat capacity × temperature change

The temperature change is 118 − 25 = 93 °C

675 J = 55.0 g × $s$ × 93 °C

$$s = \frac{675 \text{ J}}{55.0 \text{ g} \times 93\ °\text{C}} = 0.13 \text{ J/g °C}$$

## Additional Problems

64. Since $X$ is a pure substance, the fact that two different solids form when electrical current is passed indicates that $X$ must be a compound.

66. Since vaporized water is still the *same substance* as solid water, no chemical reaction has occurred. Sublimation is a physical change.

68. 2.5 kg of water = 2,500 g

Temperature change = 55.0 − 18.5 = 36.5 °C

2500 g × 4.184 J/g °C × 36.5 °C = 3.8 × $10^5$ J

70. No calculation is necessary: aluminum will lose more heat because it has the higher specific heat capacity.

72. For any substance, $Q = m \times s \times \Delta T$. The quantity of heat gained by the water in this experiment must equal the total heat lost by the metals (i.e., the *sum* of the amount of heat lost by the iron and the amount of heat lost by the aluminum).

$$(m \times s \times \Delta T)_{\text{water}} = (m \times s \times \Delta T)_{\text{iron}} + (m \times s \times \Delta T)_{\text{aluminum}}$$

If $T_f$ represents the final temperature reached by this system, then

$$[m \times s \times (T_f - 22.5\ °\text{C})]_{\text{water}}$$
$$= [m \times s \times (100\ °\text{C} - T_f)]_{\text{iron}} + [m \times s \times (100\ °\text{C} - T_f)]_{\text{aluminum}}$$

$$[97.3 \text{ g} \times 4.184 \text{ J/g °C} \times (T_f - 22.5\ °\text{C})]$$
$$= [10.00 \text{ g} \times 0.24 \text{ J/g °C} \times (100\ °\text{C} - T_f)]$$
$$+ [5.00 \text{ g} \times 0.89 \text{ J/g °C} \times (100\ °\text{C} - T_f)]$$

$$407.10(T_f - 22.5) = 4.50(100 - T_f) + 4.45(100 - T_f) = 9.95(100 - T_f)$$

$$407.10T_f - 9159.8 = 995 - 9.95T_f$$

$417.05T_f = 10{,}154.8$ which gives $T_f = 24.3$ °C

74. $(m \times s \times \Delta T)_{\text{water}} = (m \times s \times \Delta T)_{\text{iron}}$

Let $T_f$ represent the final temperature reached by the system

$[75 \text{ g} \times 4.184 \text{ J/g °C} \times (T_f - 20.)] = [25.0 \text{ g} \times 0.45 \text{ J/g °C} \times (85 - T_f)]$

$313.8\ (T_f - 20.) = 11.25\ (85 - T_f)$
$313.8T_f - 6276 = 956.25 - 11.25T_f$

$325.05T_f = 7232.25$

$T_f = 22.2 \text{ °C} = 22 \text{ °C}$

# Chapter Four Chemical Foundations: Elements and Atoms

## 4.1 The Elements

2. The chief use of gold in ancient times was as *ornamentation*, whether in statuary or in jewelry. Gold possesses an especially beautiful lustre, and since it is relatively soft and malleable, it could be worked finely by artisans; among the metals, gold is particularly inert to attack by most substances in the environment.

4. Boyle defined a substance as an element if it could not be broken down into simpler substances by chemical means.

6. The four most abundant elements in living creatures are, respectively, oxygen, carbon, hydrogen, and nitrogen (see Table 4.2). In the nonliving world, the most abundant elements are, respectively, oxygen, silicon, aluminum, and iron (see Table 4.1).

## 4.2 Symbols for the Elements

8. Sb (antimony)

   Cu (copper)

   Au (gold)

   Pb (lead)

   Hg (mercury)

   K (potassium)

   Ag (silver)

   Na (sodium)

   Sn (tin)

   W (tungsten)

10. a. Ba

    b. Br

    c. Bi

    d. B

    e. K

    f. P

12. a. Cr
    b. Cd
    c. I
    d. Cl
    e. Pt

14. a. magnesium
    b. manganese
    c. neon
    d. nickel
    e. titanium
    f. lead
    g. tungsten
    h. helium

## 4.3 Dalton's Atomic Theory

16. a. False; According to Dalton, the atoms of a given element are always *different* than the atoms of any other element.

    b. False; Atoms are *indivisible* during chemical reactions.

    c. False; Dalton's theory was *not* accepted generally for many years.

## 4.4 Formulas of Compounds

18. According to Dalton, all atoms of the same element were *identical*; in particular, every atom of a given element has the same *mass* as every other atom of that element. If a given compound always contains the *same relative numbers* of atoms of each kind, and those atoms always have the *same masses*, then it follows that the compound made from those elements would always contain the same relative masses of its elements.

20. a. $CO_2$
    b. $AlCl_3$
    c. $HClO_4$
    d. $SCl_6$
    e. $Al_2O_3$
    f. $NaN_3$

## 4.5 The Structure of the Atom

22. a. False; Rutherford's bombardment experiments with metal foil suggested that the alpha particles were being deflected by coming near a *dense*, *positively charged* atomic nucleus.

    b. False; The proton and the electron have opposite charges, but the mass of the electron is *much smaller* than the mass of the proton.

    c. True

## 4.6 Introduction to the Modern Concept of Atomic Structure

24. neutrons

26. neutron; electron

28. electrons

## 4.7 Isotopes

30. False; the *atomic* number represents the number of protons in a nucleus

32. mass

34. Atoms of the same element (i.e., atoms with the same number of protons in the nucleus) may have different numbers of neutrons, and so will have different masses.

36. a. nitrogen, N

    b. neon, Ne

    c. sodium, Na

    d. nickel, Ni

    e. titanium, Ti

    f. argon, Ar

    g. krypton, Kr

    h. xenon, Xe

38. a. $^{13}_{6}C$

    b. $^{13}_{6}C$

    c. $^{13}_{6}C$

d. $^{44}_{19}K$

e. $^{41}_{20}Ca$

f. $^{35}_{19}K$

40. a. 22 protons, 19 neutrons, 22 electrons
    b. 30 protons, 34 neutrons, 30 electrons
    c. 32 protons, 44 neutrons, 32 electrons
    d. 36 protons, 50 neutrons, 36 electrons
    e. 33 protons, 42 neutrons, 33 electrons
    f. 19 protons, 22 neutrons, 19 electrons

42.

| | | | | |
|---|---|---|---|---|
| a. | $^{40}_{20}Ca$ | 20 | 21 | 41 |
| b. | $^{55}_{25}Mn$ | 25 | 30 | 55 |
| c. | $^{109}_{47}Ag$ | 47 | 62 | 109 |
| d. | $^{45}_{21}Sc$ | 21 | 24 | 45 |

## 4.8 Introduction to the Periodic Table

44. False; The *vertical* columns in the periodic table are referred to as groups or families.

46. Metallic elements are found towards the *left* and *bottom* of the periodic table; there are far more metallic elements than there are nonmetals.

48. hydrogen, nitrogen, oxygen, fluorine, chlorine, plus all the group 8 elements (noble gases)

50. metalloids or semimetals

52. a. 8; noble gases
    b. 7; halogens
    c. 1; alkali metals
    d. 3

e. 1; alkali metals

f. 2; alkaline earth elements

g. 8; noble gases

h. 7; halogens

54. a. C; Z = 6; nonmetal

b. Se; Z = 34; nonmetal;

c. Rn; Z = 86; nonmetal; noble gases

d. Be; Z = 4; metal; alkaline earth elements

**Additional Problems**

56. a. 6

b. 1; alkali metals

c. transition metals

d. 2; alkaline earth elements

e. transition metals

f. transition metals

g. 8; noble gases

h. 2; alkaline earth elements

58.

| | element | symbol | atomic number |
|---|---|---|---|
| Group 3 | boron | B | 5 |
| | aluminum | Al | 13 |
| | gallium | Ga | 31 |
| | indium | In | 49 |
| Group 5 | nitrogen | N | 7 |
| | phosphorus | P | 15 |
| | arsenic | As | 33 |
| | antimony | Sb | 51 |
| Group 7 | helium | He | 2 |
| | neon | Ne | 10 |
| | argon | Ar | 18 |
| | krypton | Kr | 36 |

60. Most of the mass of an atom is concentrated in the nucleus: the *protons* and *neutrons* which constitute the nucleus have similar masses, and these particles are nearly two thousand times heavier than electrons. The chemical properties of an atom depend on the number and location of the *electrons* it possesses. Electrons are found in the outer regions of the atom, and are the particles most likely to be involved in interactions between atoms.

62. $C_6H_{12}O_6$

64. a. 29 protons; 34 neutrons; 29 electrons

    b. 35 protons; 45 neutrons; 35 protons

    c. 12 protons; 12 neutrons; 12 electrons

## Chapter Five    Elements, Ions, and Nomenclature

### 5.1 Natural States of the Elements

2. unreactive

4. helium

6. single atoms

8. two

10. diamond

### 5.2 Ions

12. electrons

14. 2− (two minus)

16. *-ide*

18. nonmetallic

20.

| | |
|---|---|
| [ 1] | e |
| [ 2] | a |
| [ 3] | a |
| [ 4] | g |
| [ 5] | g |
| [ 6] | f |
| [ 7] | g |
| [ 8] | a |
| [ 9] | e |
| [10] | j |

22. a. Mn (manganese), atomic number 25
$Mn^{2+}$ (25 protons, 23 electrons); Mn (25 protons, 25 electrons)

b. Ni (nickel), atomic number 28
$Ni^{2+}$ (28 protons, 26 electrons); Ni (28 protons, 28 electrons)

c. N (nitrogen), atomic number 7
$N^{3-}$ (7 protons, 10 electrons); N (7 protons, 7 electrons)

d. Co (cobalt), atomic number 27
$Co^{3+}$ (27 protons, 24 electrons); Co (27 protons, 27 electrons)

e. Fe (iron), atomic number 26
$Fe^{2+}$ (26 protons, 24 electrons); Fe (26 protons, 26 electrons)

f. P (phosphorus), atomic number 15
$P^{3-}$ (15 protons, 18 electrons); P (15 protons, 15 electrons)

24. a. none likely (element 36, Kr, is a noble gas)

b. $Ga^{3+}$ (element 31, Ga, is in Group 3)

c. $Te^{2-}$ (element 52, Te, is in Group 6)

d. $Tl^{3+}$ (element 81, Tl, is in Group 3)

e. $Br^{-}$ (element 35, Br, is in Group 7)

f. $Fr^{+}$ (element 87, Fr, is in Group 1)

## 5.3 Compounds That Contain Ions

26. Sodium chloride is an *ionic* compound, consisting of $Na^{+}$ and $Cl^{-}$ *ions*. When NaCl is dissolved in water, these ions are *set free*, and can move independently to conduct the electrical current. Sugar crystals, although they may *appear* similar visually contain *no* ions. When sugar is dissolved in water, it dissolves as uncharged *molecules*. There are no electrically charged species present in a sugar solution to carry the electrical current.

28. The total number of positive charges must equal the total number of negative charges so that there will be *no net charge* on the crystals of an ionic compound. A macroscopic sample of compound must ordinarily not have any net charge.

30. a. Two 1+ ions are needed to balance a 2− ion, so the formula must have two $Na^{+}$ ions for each $S^{2-}$ ion: $Na_2S$.

b. One 1+ ion exactly balances a 1- ion, so the formula should have an equal number of $K^{+}$ and $Cl^{-}$ ions: KCl.

c. One 2+ ion exactly balances a 2− ion, so the formula must have an equal number of $Ba^{2+}$ and $O^{2-}$ ions: BaO.

d. One 2+ ion exactly balances a 2− ion, so the formula must have an equal number of $Mg^{2+}$ and $Se^{2-}$ ions: MgSe.

e. One 2+ ion requires two 1− ions to balance charge, so the formula must have twice as many $Br^{-}$ ions as $Cu^{2+}$ ions: $CuBr_2$.

f. One 3+ ion requires three 1− ions to balance charge, so the formula must have three times as many $I^{-}$ ions as $Al^{3+}$ ions: $AlI_3$.

g. There is no simple combination of 3+ ions and 2− ions that can balance charge (try some combinations of these ions to convince yourself of this). When this happens, find the *smallest whole number* that both charges can divide into: for 3 and 2, the smallest whole number divisible by each is 6. To balance charge, we will need as many 3+ ions as it takes to give a 6+ total charge for the positive ions, and as many 2− ions as it takes to give a 6− total charge for the negative ions. Two 3+ ions give a total of

6+, whereas three 2− ions will give a total of 6−. The formula then should contain two $Al^{3+}$ ions and three $O^{2-}$ ions: $Al_2O_3$.

h. As in part g above, we must find the smallest whole number that both charges can divide into. Three 2+ ions are required to balance two 3− ions, so the formula must contain three $Ca^{2+}$ ions for every two $N^{3-}$ ions: $Ca_3N_2$.

## 5.5 Naming Compounds That Contain a Metal and a Nonmetal

32. positive ion (cation)

34. $Na^+$; $Cl^-$

36. Roman numeral (e.g., $FeCl_3$ is iron(III) chloride)

38. a. beryllium oxide
 b. magnesium iodide
 c. sodium sulfide
 d. aluminum oxide
 e. hydrogen chloride (gaseous); hydrochloric acid (aqueous)
 f. lithium fluoride
 g. silver(I) sulfide; usually called silver sulfide
 h. calcium hydride

40. a. correct

 b. Since $AlH_3$ is ionic, the prefix *tri-* is not needed.

 c. correct

 d. The hydroxide ion ($OH^-$) has a 1− charge, so iron must be in the 2+ state in this compound; the correct name is iron(II) hydroxide.

 e. The chloride ion ($Cl^-$) has a 1− charge, so cobalt must be in the 3+ state in this compound; the correct name is cobalt(III) chloride.

42. a. Since the negative ion must have a 1− charge, the iron ion must be in the 2+ state: the name is iron(II) bromide.

 b. Since sulfide ion always has a 2− charge, the cobalt ion must be in the 2+ state: the name is cobalt(II) sulfide.

 c. Since sulfide ion always has a 2− charge, and since there are three sulfide ions present, each cobalt ion must be in the 3+ state: the name is cobalt(III) sulfide.

d. Since oxide ion always has a 2− charge, the tin ion must be in the 4+ state: the name is tin(IV) oxide.

e. Since chloride ion always has a 1− charge, each mercury ion must be in the 1+ state: the name is mercury(I) chloride.

f. Since chloride ion always has a 1− charge, the mercury ion must be in the 2+ state: the name is mercury(II) chloride.

44. a. Since bromide ions always have a 1− charge, the cobalt ion must have a 3+ charge: the name is cobalt*ic* bromide.

b. Since iodide ions always have a 1− charge, the lead ion must have a 4+ charge: the name is plumb*ic* iodide.

c. Since oxide ions always have a 2− charge, and since there are three oxide ions, each iron ion must have a 3+ charge: the name is ferr*ic* oxide.

d. Since sulfide ions always have a 2− charge, the iron ion must have a 2+ charge: the name is ferr*ous* sulfide.

e. Since chloride ions always have a 1− charge, the tin ion must have a 4+ charge: the name is stann*ic* chloride.

f. Since oxide ions always have a 2− charge, the tin ion must have a 2+ charge: the name is stann*ous* oxide.

## 5.6 Naming Binary Compounds That Contain Only Nonmetals

46. a. xenon hexafluoride

b. oxygen difluoride

c. arsenic triiodide

d. dinitrogen tetraoxide (tetroxide)

e. dichlorine monoxide

f. sulfur hexafluoride

## 5.7 Naming Binary Compounds: A Review

48. a. iron(III) sulfide, ferric sulfide - ionic

b. gold(III) chloride, auric chloride - ionic

c. arsenic trihydride (arsine) - nonionic

d. chlorine monofluoride - nonionic

e. potassium oxide - ionic

f. carbon dioxide - nonionic

50. a. aluminum oxide - ionic

b. diboron trioxide - nonionic

Although boron is in Group 3, it most commonly behaves as a nonmetal forming nonionic compounds (for example, the melting point of $B_2O_3$ is only 45°C, which is much lower than the typical melting points shown by truly ionic compounds). Special properties of the boron atom which give rise to this effect will be discussed in a later chapter.

c. dinitrogen tetroxide - nonionic

d. cobalt(III) sulfide, cobaltic sulfide - ionic

e. dinitrogen pentoxide - nonionic

f. aluminum sulfide - ionic

## 5.8 Naming Compounds That Contain Polyatomic Ions

52. oxygen

54. nitrate (the ending *-ate* always implies the larger number of oxygens)

56. hypobromite

$IO_3^-$

periodate

$OI^-$

58. a. $NO_3^-$

b. $NO_2^-$

c. $NH_4^+$

d. $CN^-$

60. a. $CO_3^{2-}$

b. $HCO_3^-$

c. $C_2H_3O_2^-$

d. $CN^-$

62. a. carbonate
    b. chlorate
    c. sulfate
    d. phosphate
    e. perchlorate
    f. permanganate

64. a. lithium dihydrogen phosphate
    b. copper(II) cyanide
    c. lead(II) nitrate
    d. sodium hydrogen phosphate
    e. sodium chlorite
    f. cobalt(III) sulfate

## 5.9 Naming Acids

66. oxygen

68. a. perchloric acid
    b. iodic acid
    c. bromous acid
    d. hypochlorous acid
    e. sulfurous acid
    f. hydrocyanic acid
    g. hydrosulfuric acid
    h. phosphoric acid

## 5.10 Writing Formulas from Names

70. a. $CaCl_2$
    b. $Ag_2O$
    c. $Al_2S_3$
    d. $BeBr_2$
    e. $H_2S$
    f. KH
    g. $MgI_2$
    h. CsF

72. a. $SO_2$
    b. $N_2O$
    c. $XeF_4$
    d. $P_4O_{10}$
    e. $PCl_5$
    f. $SF_6$
    g. $NO_2$

74. a. $AgClO_4$
    b. $Co(OH)_3$
    c. $NaClO$ or $NaOCl$
    d. $K_2Cr_2O_7$
    e. $NH_4NO_2$
    f. $Fe(OH)_3$
    g. $NH_4HCO_3$
    h. $KBrO_4$

76. a. $HCN$
    b. $HNO_3$
    c. $H_2SO_4$
    d. $H_3PO_4$
    e. $HClO$ or $HOCl$
    f. $HF$
    g. $HBrO_2$
    h. $HBr$

## Additional Problems

78. Group 1: $M \rightarrow M^+ + e^-$
    Group 2: $M \rightarrow M^{2+} + 2e^-$
    Group 6: $Y + 2e^- \rightarrow Y^{2-}$
    Group 7: $X + e^- \rightarrow X^-$

80. A moist paste of NaCl would contain $Na^+$ and $Cl^-$ ions in solution, and would serve as a *conductor* of electrical impulses.

82. $H \rightarrow H^+$(hydrogen ion) $+ e^-$; $H + e^- \rightarrow H^-$(hydride ion)

84. oxyanions: $IO_3^-$; $ClO_2^-$. oxyacids: $HClO_4$; HClO; $HBrO_2$

86. a. gold(III) bromide, auric bromide
    b. cobalt(III) cyanide, cobaltic cyanide
    c. magnesium hydrogen phosphate
    d. diboron hexahydride (diborane is its common name)
    e. ammonia
    f. silver(I) sulfate (usually called silver sulfate)
    g. beryllium hydroxide

88. a. ammonium carbonate
    b. ammonium hydrogen carbonate, ammonium bicarbonate
    c. calcium phosphate
    d. sulfurous acid
    e. manganese(IV) oxide
    f. iodic acid
    g. potassium hydride

90. a. $MCl_4$
    b. $M(NO_3)_4$
    c. $MO_2$
    d. $M_3(PO_4)_4$
    e. $M(CN)_4$
    f. $M(SO_4)_2$
    g. $M(Cr_2O_7)_2$

92. $M^+$ compounds: MD, $M_2E$, $M_3F$
    $M^{2+}$ compounds: $MD_2$, ME, $M_3F_2$
    $M^{3+}$ compounds: $MD_3$, $M_2E_3$, MF

94.

| | | | | | |
|---|---|---|---|---|---|
| $CaBr_2$ | $Ca(HCO_3)_2$ | $CaH_2$ | $Ca(C_2H_3O_2)_2$ | $Ca(HSO_4)_2$ | $Ca_3(PO_4)_2$ |
| $SrBr_2$ | $Sr(HCO_3)_2$ | $SrH_2$ | $Sr(C_2H_3O_2)_2$ | $Sr(HSO_4)_2$ | $Sr_3(PO_4)_2$ |
| $NH_4Br$ | $NH_4HCO_3$ | $NH_4H$ | $NH_4C_2H_3O_2$ | $NH_4HSO_4$ | $(NH_4)_3PO_4$ |
| $AlBr_3$ | $Al(HCO_3)_3$ | $AlH_3$ | $Al(C_2H_3O_2)_3$ | $Al(HSO_4)_3$ | $AlPO_4$ |
| $FeBr_3$ | $Fe(HCO_3)_3$ | $FeH_3$ | $Fe(C_2H_3O_2)_3$ | $Fe(HSO_4)_3$ | $FePO_4$ |
| $NiBr_2$ | $Ni(HCO_3)_2$ | $NiH_2$ | $Ni(C_2H_3O_2)_2$ | $Ni(HSO_4)_2$ | $Ni_3(PO_4)_2$ |
| $AgBr$ | $AgHCO_3$ | $AgH$ | $AgC_2H_3O_2$ | $AgHSO_4$ | $Ag_3PO_4$ |
| $AuBr_3$ | $Au(HCO_3)_3$ | $AuH_3$ | $Au(C_2H_3O_2)_3$ | $Au(HSO_4)_3$ | $AuPO_4$ |
| $KBr$ | $KHCO_3$ | $KH$ | $KC_2H_3O_2$ | $KHSO_4$ | $K_3PO_4$ |
| $HgBr_2$ | $Hg(HCO_3)_2$ | $HgH_2$ | $Hg(C_2H_3O_2)_2$ | $Hg(HSO_4)_2$ | $Hg_3(PO_4)_2$ |
| $BaBr_2$ | $Ba(HCO_3)_2$ | $BaH_2$ | $Ba(C_2H_3O_2)_2$ | $Ba(HSO_4)_2$ | $Ba_3(PO_4)_2$ |

## Chapter Six Chemical Reactions: An Introduction

### 6.1 Evidence for a Chemical Reaction

2. *Heat* is evolved as drain cleaners work (often boiling any water in the clogged drain). Some drain cleaners containing small shavings of magnesium also *bubble* (hydrogen gas is formed) as the reaction takes place.

4. Most oils *decompose* at high temperatures, producing a foul-smelling chemical called acrolein that we associate with the smell and taste of burned food. Oil that has been heated to too high a temperature for too long a period "turns rancid" and is unfit for further use. Also, at high enough temperatures, the oil might ignite.

6. Many over-the-counter antacids contain either carbonate ion ($CO_3^{2-}$) or hydrogen carbonate ion ($HCO_3^-$). When either of these encounter stomach acid (primarily HCl), carbon dioxide gas is released.

### 6.2 Chemical Equations

8. atoms

10. the same

12. water

14. $Fe(s) + S(s) \rightarrow FeS(s)$

16. $Na(s) + Cl_2(g) \rightarrow NaCl(s)$

18. $C_3H_8(g) + O_2(g) \rightarrow CO_2(g) + H_2O(g)$

20. $H_2S(g) + Pb(NO_3)_2(aq) \rightarrow PbS(s) + HNO_3(aq)$

22. $KI(aq) + H_2O(l) \rightarrow KOH(aq) + H_2(g) + I_2(s)$

24. $Mg(s) + O_2(g) \rightarrow MgO(s)$

26. $Mg(s) + H_2O(g) \rightarrow Mg(OH)_2(s) + H_2(g)$

28. $P_4(s) + O_2(g) \rightarrow P_4O_{10}(s)$

30. $CuO(s) + H_2SO_4(aq) \rightarrow CuSO_4(aq) + H_2O(l)$

32. $PbS(s) + O_2(g) \rightarrow PbO(s) + SO_2(g)$

34. $Na_2SO_3(aq) + S(s) \rightarrow Na_2S_2O_3(aq)$

## 6.3 Balancing Chemical Equations

36. whole numbers (integers)

38. For simplicity, the physical states of the substances have been omitted until the final balanced equation is obtained.

a. $Cl_2 + KBr \rightarrow Br_2 + KCl$

Balance chlorine: $Cl_2 + KBr \rightarrow Br_2 + 2KCl$

Balance bromine: $Cl_2 + 2KBr \rightarrow Br_2 + 2KCl$

Balanced equation: $Cl_2(g) + 2KBr(aq) \rightarrow Br_2(l) + 2KCl(aq)$

b. $Cr + O_2 \rightarrow Cr_2O_3$

Balance oxygen: We need to get three oxygen atoms for each $Cr_2O_3$, but the source of oxygen is $O_2$; however, three $O_2$ molecules would provide six oxygen atoms, which would be the number of oxygen atoms needed for two $Cr_2O_3$.

$Cr + 3O_2 \rightarrow 2Cr_2O_3$

Balance chromium: $4Cr + 3O_2 \rightarrow 2Cr_2O_3$

Balanced equation: $4Cr(s) + 3O_2(g) \rightarrow 2Cr_2O_3(s)$

c. $P_4 + H_2 \rightarrow PH_3$

Balance phosphorus: $P_4 + H_2 \rightarrow 4PH_3$

Balance hydrogen: $P_4 + 6H_2 \rightarrow 4PH_3$

Balanced equation: $P_4(s) + 6H_2(g) \rightarrow 4PH_3(g)$

d. $Al + H_2SO_4 \rightarrow Al_2(SO_4)_3 + H_2$

Hint: since sulfate ions are present on each side of the equation [in $H_2SO_4$ and in $Al_2(SO_4)_3$], and do not break down during the reaction, balance sulfate ions as a single unit (not as individual sulfur and oxygen atoms).

Balance sulfate ions: $Al + 3H_2SO_4 \rightarrow Al_2(SO_4)_3 + H_2$

Balance hydrogen: $Al + 3H_2SO_4 \rightarrow Al_2(SO_4)_3 + 3H_2$

Balance aluminum: $2Al + 3H_2SO_4 \rightarrow Al_2(SO_4)_3 + 3H_2$

Balanced equation: $2Al(s) + 3H_2SO_4(aq) \rightarrow Al_2(SO_4)_3(aq) + 3H_2(g)$

e. $PCl_3 + H_2O \rightarrow H_3PO_3 + HCl$

Balance chlorine: $PCl_3 + H_2O \rightarrow H_3PO_3 + 3HCl$

Balance hydrogen: $PCl_3 + 3H_2O \rightarrow H_3PO_3 + 3HCl$

Oxygen has also now been balanced.

Balanced equation: $PCl_3(l) + 3H_2O(l) \rightarrow H_3PO_3(aq) + 3HCl(aq)$

f. $SO_2 + O_2 \rightarrow SO_3$

This equation looks so simple, but can be hard to balance by inspection. The left side of the equation has four oxygen atoms, while the right side has only three oxygen atoms. If we had *individual* oxygen atoms available, balancing would be simple, but oxygen occurs naturally as $O_2$ molecules, which are all we can use to balance oxygen atoms (you cannot change the formulas of any of the reactants or products when balancing an equation). If we put a coefficient of 2 in front of each of the sulfur compounds, however, we go from having a difference of *one* oxygen atom between the sides of the equation to having a difference of *two* oxygen atoms, which makes for one $O_2$ molecule: $2SO_2 + O_2 \rightarrow 2SO_3$

Balanced equation: $2SO_2(g) + O_2(g) \rightarrow 2SO_3(g)$

g. $C_7H_{16} + O_2 \rightarrow CO_2 + H_2O$

Balance carbon: $C_7H_{16} + O_2 \rightarrow 7CO_2 + H_2O$

Balance hydrogen: $C_7H_{16} + O_2 \rightarrow 7CO_2 + 8H_2O$

Balance oxygen: $C_7H_{16} + 11O_2 \rightarrow 7CO_2 + 8H_2O$

Balanced equation: $C_7H_{16}(l) + 11O_2(g) \rightarrow 7CO_2(g) + 8H_2O(g)$

h. $C_2H_6 + O_2 \rightarrow CO_2 + H_2O$

Hint: equations involving the reaction with $O_2$ of carbon/hydrogen compounds in which there is an *even* number of carbon atoms very often require a coefficient of 2 for the carbon/hydrogen compound in order to balance the number of oxygen atoms used.

Balance hydrogen: $C_2H_6 + O_2 \rightarrow CO_2 + 3H_2O$

Balance carbon: $C_2H_6 + O_2 \rightarrow 2CO_2 + 3H_2O$

Notice that seven oxygen atoms are needed to balance the number of oxygen atoms on the right side of the equation, but that oxygen occurs on the left side of the equation as $O_2$ molecules, which cannot be changed. However, if a coefficient of 2 is placed in front of $C_2H_6$, and carbon and hydrogen are *re*balanced, a whole number of $O_2$ molecules will be needed.

Add coefficient of 2 for $C_2H_6$: $2C_2H_6 + O_2 \rightarrow 2CO_2 + 3H_2O$

Rebalance carbon: $2C_2H_6 + O_2 \rightarrow 4CO_2 + 3H_2O$

Rebalance hydrogen $2C_2H_6 + O_2 \rightarrow 4CO_2 + 6H_2O$

Balance oxygen: $2C_2H_6 + 7O_2 \rightarrow 4CO_2 + 6H_2O$

Balanced equation: $2C_2H_6(g) + 7O_2(g) \rightarrow 4CO_2(g) + 6H_2O(g)$

40. For simplicity, the physical states of the substances have been omitted until the final balanced equation is obtained.

a. $Ba(NO_3)_2 + KF \rightarrow BaF_2 + KNO_3$

Balance nitrate ion: $Ba(NO_3)_2 + KF \rightarrow BaF_2 + 2KNO_3$

Balance potassium: $Ba(NO_3)_2 + 2KF \rightarrow BaF_2 + 2KNO_3$

Fluorine and barium are now balanced also.

Balanced equation: $Ba(NO_3)_2(aq) + 2KF(aq) \rightarrow BaF_2(s) + 2KNO_3(aq)$

b. $Zn + HCl \rightarrow ZnCl_2 + H_2$

Balance hydrogen: $Zn + 2HCl \rightarrow ZnCl_2 + H_2$

Chlorine is also now balanced.

Balanced equation: $Zn(s) + 2HCl(aq) \rightarrow ZnCl_2(aq) + H_2(g)$

c. $Fe + S \rightarrow Fe_2S_3$

Balance Fe: $2Fe + S \rightarrow Fe_2S_3$

Balance S: $2Fe + 3S \rightarrow Fe_2S_3$

Balanced equation: $2Fe(s) + 3S(s) \rightarrow Fe_2S_3(s)$

d. $C_6H_{12}O_6 + O_2 \rightarrow CO_2 + H_2O$

Balance carbon: $C_6H_{12}O_6 + O_2 \rightarrow 6CO_2 + H_2O$

Balance hydrogen: $C_6H_{12}O_6 + O_2 \rightarrow 6CO_2 + 6H_2O$

Balance oxygen: $C_6H_{12}O_6 + 6O_2 \rightarrow 6CO_2 + 6H_2O$

Balanced equation: $C_6H_{12}O_6(s) + 6O_2(g) \rightarrow 6CO_2(g) + 6H_2O(g)$

e. $H_2O + Cl_2 \rightarrow HCl + HOCl$

Don't get caught! This equation is already balanced. Always count atoms before beginning to balance.

f. $ZnS + O_2 \rightarrow ZnO + SO_2$

Notice that the right side of the equation has *three* oxygen atoms, whereas the left side has only *two* oxygen atoms. The only source of oxygen atoms on the left side of the equation is as $O_2$ molecules. We will have to end up with an *even number* of oxygen atoms on the right side of the equation, since we cannot change the formula of $O_2$. So try adding a coefficient of 2 for ZnO to make an even number of oxygen atoms on the right side

$ZnS + O_2 \rightarrow 2ZnO + SO_2$

Balance zinc: $2ZnS + O_2 \rightarrow 2ZnO + SO_2$

Balance sulfur: $2ZnS + O_2 \rightarrow 2ZnO + 2SO_2$

Balance oxygen: $2ZnS + 3O_2 \rightarrow 2ZnO + 2SO_2$

Balanced equation: $2ZnS(s) + 3O_2(g) \rightarrow 2ZnO(s) + 2SO_2(g)$

g. $PbSO_4 + NaCl \rightarrow Na_2SO_4 + Na_2PbCl_4$

Balance chlorine: $PbSO_4 + 4NaCl \rightarrow Na_2SO_4 + Na_2PbCl_4$

All other atoms are now balanced.

Balanced equation: $PbSO_4(s) + 4NaCl(aq) \rightarrow Na_2SO_4(aq) + Na_2PbCl_4(aq)$

h. $Fe_2O_3 + C \rightarrow Fe_3O_4 + CO$

Notice that we have two iron atoms on the *left* side of the equation and three iron atoms on the *right* side. There is no simple multiple of two or three that will give the right number of iron atoms for the opposite side of the equation. The smallest whole number that is divisible by both 2 and 3 is *six*, so the coefficients should be adjusted so that there are six iron atoms on each side of the equation

Balance iron: $3Fe_2O_3 + C \rightarrow 2Fe_3O_4 + CO$

All other atoms are now balanced.

Balanced equation: $3Fe_2O_3(s) + C(s) \rightarrow 2Fe_3O_4(s) + CO(g)$

42. a. $SiCl_4(l) + 2Mg(s) \rightarrow Si(s) + 2MgCl_2(s)$

b. $2NO(g) + Cl_2(g) \rightarrow 2NOCl(g)$

c. $3MnO_2(s) + 4Al(s) \rightarrow 3Mn(s) + 2Al_2O_3(s)$

d. $16Cr(s) + 3S_8(s) \rightarrow 8Cr_2S_3(s)$

e. $4NH_3(g) + 3F_2(g) \rightarrow 3NH_4F(s) + NF_3(g)$

f. $Ag_2S(s) + H_2(g) \rightarrow 2Ag(s) + H_2S(g)$

g. $3O_2(g) \rightarrow 2O_3(g)$

h. $8Na_2SO_3(aq) + S_8(s) \rightarrow 8Na_2S_2O_3(aq)$

44. a. $Pb(NO_3)_2(aq) + K_2CrO_4(aq) \rightarrow PbCrO_4(s) + 2KNO_3(aq)$

b. $BaCl_2(aq) + Na_2SO_4(aq) \rightarrow BaSO_4(s) + 2NaCl(aq)$

c. $2CH_3OH(l) + 3O_2(g) \rightarrow 2CO_2(g) + 4H_2O(g)$

d. $Na_2CO_3(aq) + S(s) + SO_2(g) \rightarrow CO_2(g) + Na_2S_2O_3(aq)$

e. $Cu(s) + 2H_2SO_4(aq) \rightarrow CuSO_4(aq) + SO_2(g) + 2H_2O(l)$

f. $MnO_2(s) + 4HCl(aq) \rightarrow MnCl_2(aq) + Cl_2(g) + 2H_2O(l)$

g. $As_2O_3(s) + 6KI(aq) + 6HCl(aq) \rightarrow 2AsI_3(s) + 6KCl(aq) + 3H_2O(l)$

h. $2Na_2S_2O_3(aq) + I_2(aq) \rightarrow Na_2S_4O_6(aq) + 2NaI(aq)$

## Additional Problems

46. The beauty of the periodic chart is, if you know how *one* element of a group reacts, then you can be almost certain that all the other members of the group will react in a similar fashion. This permits you to write a "group reaction" that summarizes the properties of the entire group. For example, if M represents a typical alkali metal (Group 1), then the unbalanced reaction of M with water can be written as

$M + H_2O \rightarrow MOH + H_2$

This reaction applies for all the metals of Group 1:

$Li(s) + H_2O(l) \rightarrow LiOH(aq) + H_2(g)$

$Na(s) + H_2O(l) \rightarrow NaOH(aq) + H_2(g)$

$K(s) + H_2O(l) \rightarrow KOH(aq) + H_2(g)$

$Rb(s) + H_2O(l) \rightarrow RbOH(aq) + H_2(g)$

$Cs(s) + H_2O(l) \rightarrow CsOH(aq) + H_2(g)$

$Fr(s) + H_2O(l) \rightarrow FrOH(aq) + H_2(g)$

Notice the correspondence between these specific reactions and the general group reaction for M.

48. $C_{12}H_{22}O_{11}(aq) + H_2O(l) \rightarrow 4C_2H_6O(aq) + 4CO_2(g)$

50. $2Al_2O_3(s) + 3C(s) \rightarrow 4Al(s) + 3CO_2(g)$

52. Let M represent a Group 2 metal, and $X_2$ represent the halogens; then

$M + X_2 \rightarrow MX_2$

represents the general group reaction. The specific reactions are

$Be(s) + F_2(g) \rightarrow BeF_2(s)$

$Be(s) + Cl_2(g) \rightarrow BeCl_2(s)$

$Mg(s) + F_2(g) \rightarrow MgF_2(s)$

$Mg(s) + Cl_2(g) \rightarrow MgCl_2(s)$

$Ca(s) + F_2(g) \rightarrow CaF_2(s)$

$Ca(s) + Cl_2(g) \rightarrow CaCl_2(s)$

$Sr(s) + F_2(g) \rightarrow SrF_2(s)$

$Sr(s) + Cl_2(g) \rightarrow SrCl_2(s)$

$Ba(s) + F_2(g) \rightarrow BaF_2(s)$

$Ba(s) + Cl_2(g) \rightarrow BaCl_2(s)$

$Ra(s) + F_2(g) \rightarrow RaF_2(s)$

$Ra(s) + Cl_2(g) \rightarrow RaCl_2(s)$

54. $(NH_4)_2Cr_2O_7(s) \rightarrow Cr_2O_3(s) + N_2(g) + 4H_2O(g)$

56. $2KClO_3(s) \rightarrow 2KCl(s) + 3O_2(g)$

58. $NH_3(g) + HCl(g) \rightarrow NH_4Cl(s)$

# Chapter Seven Reactions in Aqueous Solutions

## 7.1 Predicting Whether a Reaction Will Occur

2. Driving forces are types of *changes* in a system which pull a reaction in the *direction of product formation*; driving forces discussed in Chapter Seven include: formation of a *solid*, formation of *water*, formation of a *gas*, transfer of electrons.

## 7.2 Reactions in Which a Solid Forms

4. The net charge of a precipitate must be *zero*. The total number of positive charges equals the total number of negative charges.

6. ions

8. Consider the two new possible combinations of ions (when two ionic compounds are mixed the ions may "switch partners"); if one of these new possible combinations is insoluble in water it will form a precipitate having zero net charge (indicating in what proportions the ions must be combined). There is no foolproof method for predicting by inspection what precipitate may form: we rely on intuition, experience from similar reactions, and "rules of thumb" based on the results of experiments (such as the general solubility rules from Table 7.1).

10. For most practical purposes, "insoluble" and "slightly" soluble mean the same thing. However, if a substance were highly toxic and were found in a water supply, for example, the difference between "insoluble" and "slightly soluble" could be crucial.

12. a. soluble (Rule 2: most potassium salts are soluble)
    b. soluble (Rule 2: most ammonium salts are soluble)
    c. insoluble (Rule 6: most carbonate salts are only slightly soluble)
    d. insoluble (Rule 6: most phosphate salts are only slightly soluble)
    e. soluble (Rule 2: most sodium salts are soluble)
    f. insoluble (Rule 5: most hydroxide salts are only slightly soluble)
    g. soluble (Rule 3: most chloride salts are soluble)

14. a. Rule 6: most sulfide salts are only slightly soluble
    b. Rule 5: most hydroxide salts are only slightly soluble
    c. Rule 6: most carbonate salts are only slightly soluble
    d. Rule 6: most phosphate salts are only slightly soluble

16. a. iron(III) hydroxide, $Fe(OH)_3$. Rule 5: most hydroxide salts are only slightly soluble.

b. nickel(II) sulfide, NiS. Rule 6: most sulfide salts are only slightly soluble.

c. silver chloride, AgCl. Rule 3: Although most chloride salts are soluble, AgCl is a listed exception.

d. barium carbonate, $BaCO_3$. Rule 6: most carbonate salts are only slightly soluble.

e. mercury(I) chloride or mercurous chloride, $Hg_2Cl_2$. Rule 3: Although most chloride salts are soluble, $Hg_2Cl_2$ is a listed exception.

f. barium sulfate, $BaSO_4$. Rule 4: Although most sulfate salts are soluble, $BaSO_4$ is a listed exception

18. The precipitates are marked in boldface type.

a. Rule 1: most nitrate salts are soluble.

$AgNO_3(aq) + Ba(NO_3)_2(aq) \rightarrow$ *no* precipitate

b. Rule 3: AgCl is listed as an exception.

$NiCl_2(aq) + 2AgC_2H_3O_2(aq) \rightarrow \mathbf{AgCl}(s) + Ni(C_2H_3O_2)_2(aq)$

c. Rule 6: most carbonate salts are only slightly soluble.

$Pb(NO_3)_2(aq) + (NH_4)_2CO_3(aq) \rightarrow \mathbf{PbCO_3}(s) + 2NH_4NO_3(aq)$

d. Rule 4: $CaSO_4$ is listed as an exception.

$Na_2SO_4(aq) + Ca(C_2H_3O_2)_2(aq) \rightarrow \mathbf{CaSO_4}(s) + 2NaC_2H_3O_2(aq)$

e. Rule 3: $Hg_2Cl_2$ is listed as an exception.

$NiCl_2(aq) + Hg_2(NO_3)_2(aq) \rightarrow \mathbf{Hg_2Cl_2}(s) + Ni(NO_3)_2(aq)$

f. Rule 4: $CaSO_4$ is listed as an exception.

$Ca(NO_3)_2(aq) + H_2SO_4(aq) \rightarrow \mathbf{CaSO_4}(s) + 2HNO_3(aq)$

20. Hint: when balancing equations involving polyatomic ions, especially in precipitation reactions, balance the polyatomic ions as a *unit*, not in terms of the atoms the polyatomic ions contain (e.g., treat nitrate ion, $NO_3^-$ as a single entity, not as one nitrogen and three oxygen atoms). When finished balancing, however, do be sure to count the individual number of atoms of each type on each side of the equation.

a. $Co(NO_3)_2(aq) + (NH_4)_2S(aq) \rightarrow CoS(s) + NH_4NO_3(aq)$

Balance $NO_3^-$ ions: $Co(NO_3)_2(aq) + (NH_4)_2S(aq) \rightarrow CoS(s) + 2NH_4NO_3(aq)$

Equation is now balanced.

b. $CaCl_2(aq) + H_2SO_4(aq) \rightarrow CaSO_4(s) + HCl(aq)$

Balance chlorine: $CaCl_2(aq) + H_2SO_4(aq) \rightarrow CaSO_4(s) + 2HCl(aq)$

Equation is now balanced.

c. $FeCl_3(aq) + Na_3PO_4(aq) \rightarrow FePO_4(s) + NaCl(aq)$

Balance chlorine: $FeCl_3(aq) + Na_3PO_4(aq) \rightarrow FePO_4(s) + 3NaCl(aq)$

Equation is now balanced.

22. The products are determined by having the ions "switch partners." For example, for a general reaction AB + CD →, the possible products are AD and CB if the ions switch partners. If either AD or CB is insoluble, then a precipitation reaction has occurred. In the following reaction, the formula of the precipitate is given in boldface type.

a. $Pb(NO_3)_2(aq) + Na_2CO_3(aq) \rightarrow \mathbf{PbCO_3}(s) + 2NaNO_3(aq)$

Solubility Rule 6: most $CO_3^-$ salts are only slightly soluble

b. $K_2SO_4(aq) + CaCl_2(aq) \rightarrow \mathbf{CaSO_4}(s) + 2KCl(aq)$

Solubility Rule 4: $CaSO_4$ is listed as an insoluble exception

c. $Ni(C_2H_3O_2)_2(aq) + Na_2S(aq) \rightarrow \mathbf{NiS}(s) + 2NaC_2H_3O_2(aq)$

Solubility Rule 6: most $S^{2-}$ salts are only slightly soluble

## 7.3 Describing Reactions in Aqueous Solutions

24. spectator

26. The net ionic equation for a reaction indicates *only those ions that go to form the precipitate*, and does not show the spectator ions present in the solutes mixed. The identity of the precipitate is determined from the Solubility Rules (Table 7.1).

a. $Fe^{3+}(aq) + 3OH^-(aq) \rightarrow Fe(OH)_3(s)$

Rule 5: most hydroxide compounds are only slightly soluble

b. $Ni^{2+}(aq) + S^{2-}(aq) \rightarrow NiS(s)$

Rule 6: most sulfide salts are only slightly soluble

c. $Ag^+(aq) + Cl^-(aq) \rightarrow AgCl(s)$

Rule 3: AgCl is listed as an insoluble exception

d. $Ba^{2+}(aq) + SO_4^{2-}(aq) \rightarrow BaSO_4(s)$

Rule 4: $BaSO_4$ is listed as an insoluble exception

e. $Hg_2^{2+}(aq) + 2Br^-(aq) \rightarrow Hg_2Br_2(s)$

Rule 3: $Hg_2Cl_2$ is listed as insoluble; since Br occurs in the same group of the periodic table as Cl, $Hg_2Br_2$ would be expected to be insoluble also.

f. $Ba^{2+}(aq) + SO_4^{2-}(aq) \rightarrow BaSO_4(s)$

Rule 4: $BaSO_4$ is listed as an insoluble exception

28. $Ba^{2+}(aq) + SO_4^{2-}(aq) \rightarrow BaSO_4(s)$

30. $Fe^{2+}(aq) + S^{2-}(aq) \rightarrow FeS(s)$

$2Cr^{3+}(aq) + 3S^{2-}(aq) \rightarrow Cr_2S_3(s)$

$Ni^{2+}(aq) + S^{2-}(aq) \rightarrow NiS(s)$

## 7.4 Reactions That Form Water: Acids and Bases

32. acid

34. HCl

36. salt

38. $RbOH(s) \rightarrow Rb^+(aq) + OH^-(aq)$

$CsOH(s) \rightarrow Cs^+(aq) + OH^-(aq)$

40. The formulas of the salts are marked in boldface type. Remember that in an acid/base reaction in aqueous solution, *water* is always one of the products: keeping this in mind makes predicting the formula of the *salt* produced easy to do.

a. $HClO_4(aq) + RbOH(aq) \rightarrow H_2O(l) + \mathbf{RbClO_4}(aq)$

b. $HNO_3(aq) + KOH(aq) \rightarrow H_2O(l) + \mathbf{KNO_3}(aq)$

c. $H_2SO_4(aq) + 2NaOH(aq) \rightarrow 2H_2O(l) + \mathbf{Na_2SO_4}(aq)$

d. $HBr(aq) + CsOH(aq) \rightarrow H_2O(l) + \mathbf{CsBr}(aq)$

## Additional Problems

42. A *molecular equation* uses the normal, uncharged formulas for the compounds involved. The *complete ionic equation* shows the compounds involved broken up into their respective ions (*all* ions present are shown). The *net ionic equation* shows only those ions which combine to form a precipitate, a gas, or a nonionic product such as water. The net ionic equation shows most clearly the species that are combining with each other.

44. $Pb^{2+}(aq) + 2Cl^-(aq) \rightarrow PbCl_2(s)$

$Pb^{2+}(aq) + CrO_4^{2-}(aq) \rightarrow PbCrO_4(s)$

46. The formulas of the precipitates are predicted using the Solubility Rules of Table 7.1. You should be able by now to predict the formation of a precipitate without having to write out full molecular or ionic equations.

a. Rule 5: $Co^{3+}(aq) + 3OH^-(aq) \rightarrow Co(OH)_3(s)$

b. Rule 6: $2Ag^+(aq) + CO_3^{2-}(aq) \rightarrow Ag_2CO_3(s)$

c. no reaction (all combinations of ions are soluble)

d. Rule 4: $Ba^{2+}(aq) + SO_4^{2-}(aq) \rightarrow BaSO_4(s)$

e. no reaction (all combinations of ions are soluble)

f. Rule 6: $3Ca^{3+}(aq) + 2PO_4^{3-}(aq) \rightarrow Ca_3(PO_4)_2(s)$

g. Rule 5: $Al^{3+}(aq) + 3OH^-(aq) \rightarrow Al(OH)_3(s)$

48. $Ca(OH)_2(s) \rightarrow Ca^{2+}(aq) + 2OH^-(aq)$

$Mg(OH)_2(s) \rightarrow Mg^{2+}(aq) + 2OH^-(aq)$

$Sr(OH)_2(s) \rightarrow Sr^{2+}(aq) + 2OH^-(aq)$

$Ba(OH)_2(s) \rightarrow Ba^{2+}(aq) + 2OH^-(aq)$

50. When *any* strong acid is reacted with *any* strong base, the net ionic reaction is the *same:*

$H^+(aq) + OH^-(aq) \rightarrow H_2O(l)$

Since the net ionic reaction is the same for any strong acid, the amount of heat liberated is also the same.

# Chapter Eight Classifying Chemical Reactions

## 8.1 Reactions of Metals with Nonmetals (Oxidation-Reduction)

2. transfer

4. Metal atoms *lose* electrons and form *cations*; *non*metal atoms *gain* electrons and become *anions*.

6. Chlorine atoms would each gain one electron to become $Cl^-$ ions. Sulfur atoms would each gain two electrons to become $S^{2-}$ ions.

8. $Fe_2S_3$ is made up of $Fe^{3+}$ and $S^{2-}$ ions. Iron atoms each lose three electrons to become $Fe^{3+}$ ions. Sulfur atoms each gain two electrons to become $S^{2-}$ ions.

10. a. $Na + O_2 \rightarrow Na_2O_2$

    Balance sodium: $2Na + O_2 \rightarrow Na_2O_2$

    Balanced equation: $2Na(s) + O_2(g) \rightarrow Na_2O_2(s)$

    b. $Fe(s) + H_2SO_4(aq) \rightarrow FeSO_4(aq) + H_2(g)$

    Equation is already balanced!

    c. $Al_2O_3 \rightarrow Al + O_2$

    Balance oxygen: $2Al_2O_3 \rightarrow Al + 3O_2$

    Balance aluminum: $2Al_2O_3 \rightarrow 4Al + 3O_2$

    Balanced equation: $2Al_2O_3(s) \rightarrow 4Al(s) + 3O_2(g)$

    d. $Fe + Br_2 \rightarrow 2FeBr_3$

    Balance bromine: $Fe + 3Br_2 \rightarrow 2FeBr_3$

    Balance iron: $2Fe + 3Br_2 \rightarrow 2FeBr_3$

    Balanced equation: $2Fe(s) + 3Br_2(l) \rightarrow 2FeBr_3(s)$

    e. $Zn + HNO_3 \rightarrow Zn(NO_3)_2 + H_2$

    Balance nitrate ions: $Zn + 2HNO_3 \rightarrow Zn(NO_3)_2 + H_2$

    Balanced equation: $Zn(s) + 2HNO_3(aq) \rightarrow Zn(NO_3)_2(aq) + H_2(g)$

## 8.2 Ways to Classify Reactions

12. examples of formation of water:

    $HCl(aq) + NaOH(aq) \rightarrow H_2O(l) + NaCl(aq)$

    $H_2SO_4(aq) + 2KOH(aq) \rightarrow 2H_2O(l) + K_2SO_4(aq)$

    examples of formation of a gaseous product

    $Mg(s) + 2HCl(aq) \rightarrow MgCl_2(aq) + H_2(g)$

    $2KClO_3(s) \rightarrow 2KCl(s) + 3O_2(g)$

14. For each reaction, the type of reaction is first identified, followed by some of the reasoning that leads to this choice (there may be more than one way in which you can recognize a particular type of reaction).

    a. oxidation-reduction (Fe changes from the elemental state to the combined state in $Fe_3(SO_4)_2$; hydrogen changes from the combined to the elemental state).

    b. acid-base ($HClO_4$ is a strong acid and RbOH is a strong base; water and a salt are produced).

    c. oxidation-reduction (both Ca and $O_2$ change from the elemental to the combined state).

    d. acid-base ($H_2SO_4$ is a strong acid and NaOH is a strong base; water and a salt are produced).

    e. precipitation (from the Solubility Rules of Table 7.1, $PbCO_3$ is insoluble).

    f. precipitation (from the Solubility Rules of Table 7.1, $CaSO_4$ is insoluble).

    g. acid-base ($HNO_3$ is a strong acid and KOH is a strong base; water and a salt are produced).

    h. precipitation (from the Solubility Rules of Table 7.1, NiS is insoluble).

    i. oxidation-reduction (both Ni and $Cl_2$ change from the elemental to the combined state).

## 8.3 Other Ways to Classify Reactions

16. oxidation-reduction

18. decomposition

20. a. $C_5H_{12}(l) + 8O_2(g) \rightarrow 5CO_2(g) + 6H_2O(g)$

    b. $C_2H_6O(l) + 3O_2(g) \rightarrow 2CO_2(g) + 3H_2O(g)$

    c. $2C_6H_6(l) + 15O_2(g) \rightarrow 12CO_2(g) + 6H_2O(g)$

22. The products of the combustion of hydrocarbons in oxygen are almost always $CO_2(g)$ and $H_2O(g)$:

    a. $C_2H_4(g) + 3O_2(g) \rightarrow 2CO_2(g) + 2H_2O(g)$

    b. $2C_8H_{18}(l) + 25O_2(g) \rightarrow 16CO_2(g) + 18H_2O(g)$

    c. $2C_{30}H_{62}(s) + 91O_2(g) \rightarrow 60CO_2(g) + 62H_2O(g)$

24. a. $4FeO(s) + O_2(g) \rightarrow 2Fe_2O_3(s)$

    b. $2CO(g) + O_2(g) \rightarrow 2CO_2(g)$

    c. $H_2(g) + Cl_2(g) \rightarrow 2HCl(g)$

    d. $16K(s) + S_8(s) \rightarrow 8K_2S(s)$

    e. $6Na(s) + N_2(g) \rightarrow 2Na_3N(s)$

26. a. $2NaHCO_3(s) \rightarrow Na_2CO_3(s) + H_2O(g) + CO_2(g)$

    b. $2NaClO_3(s) \rightarrow 2NaCl(s) + 3O_2(g)$

    c. $2HgO(s) \rightarrow 2Hg(l) + O_2(g)$

    d. $C_{12}H_{22}O_{11}(s) \rightarrow 12C(s) + 11H_2O(g)$

    e. $2H_2O_2(l) \rightarrow 2H_2O(l) + O_2(g)$

## Additional Problems

28. a. two; $O + 2e^- \rightarrow O^{2-}$

    b. one; $F + e^- \rightarrow F^-$

    c. three; $N + 3e^- \rightarrow N^{3-}$

    d. one; $Cl + e^- \rightarrow Cl^-$

    e. two; $S + 2e^- \rightarrow S^{2-}$

    f. one; $Br + e^- \rightarrow Br^-$

30. carbon dioxide gas and water vapor

32. a. $2C_3H_8O(l) + 9O_2(g) \rightarrow 6CO_2(g) + 8H_2O(g)$
    oxidation-reduction, combustion

b. $HCl(aq) + AgC_2H_3O_2(aq) \rightarrow AgCl(s) + HC_2H_3O_2(aq)$

precipitation

c. $3HCl(aq) + Al(OH)_3(s) \rightarrow AlCl_3(aq) + 3H_2O(l)$

acid-base

d. $2H_2O_2(aq) \rightarrow 2H_2O(l) + O_2(g)$

oxidation-reduction, decomposition

e. $N_2H_4(l) + O_2(g) \rightarrow N_2(g) + 2H_2O(g)$

oxidation-reduction, combustion

34. $2Zn(s) + O_2(g) \rightarrow 2ZnO(s)$

$4Al(s) + 3O_2(g) \rightarrow 2Al_2O_3(s)$

$2Fe(s) + O_2(g) \rightarrow 2FeO(s)$; $4Fe(s) + 3O_2(g) \rightarrow 2Fe_2O_3(s)$

$2Cr(s) + O_2(g) \rightarrow 2CrO(s)$; $4Cr(s) + 3O_2(g) \rightarrow 2Cr_2O_3(s)$

$2Ni(s) + O_2(g) \rightarrow 2NiO(s)$

36.

| | |
|---|---|
| Al | 3+ |
| Ba | 2+ |
| Br | 1− |
| Ca | 2+ |
| Cl | 1− |
| Cs | 1+ |
| I | 1− |
| K | 1+ |
| Li | 1+ |
| Mg | 2+ |
| Na | 1+ |
| O | 2− |
| Rb | 1+ |
| S | 2− |
| Sr | 2+ |

# Chapter Nine Chemical Composition

## 9.1 Counting by Weighing

2. $500.\ \text{g} \times \dfrac{1 \text{ cork}}{1.63 \text{ g}} = 306.7 = 307 \text{ corks}$

$500.\ \text{g} \times \dfrac{1 \text{ stopper}}{4.31 \text{ g}} = 116 \text{ stoppers}$

1 Kg (1000 g) of corks contains $(1000 \text{ g} \times \dfrac{1 \text{ cork}}{1.63 \text{ g}}) = 613.49 = 613 \text{ corks}$

613 stoppers would weigh $(613 \text{ stoppers} \times \dfrac{4.31 \text{ g}}{1 \text{ stopper}}) = 2644 \text{ g} = 2640 \text{ g}$

The ratio of the mass of a stopper to the mass of a cork is (4.31 g/1.63 g). So the mass of stoppers that contains the same number of stoppers as there are corks in 1000 g of corks is

$1000 \text{ g} \times \dfrac{4.31 \text{ g}}{1.63 \text{ g}} = 2644 \text{ g} = 2640 \text{ g}$

## 9.2 Atomic Masses: Counting Atoms by Weighing

4. average

6. a. $160{,}000 \text{ amu} \times \dfrac{1 \text{ O atom}}{16.00 \text{ amu}} = 1.0 \times 10^4 \text{ O atoms}$ (assuming exact)

b. $8139.81 \text{ amu} \times \dfrac{1 \text{ N atom}}{14.01 \text{ amu}} = 581.0 \text{ N atoms}$

c. $13{,}490 \text{ amu} \times \dfrac{1 \text{ Al atom}}{26.98 \text{ amu}} = 500.0 \text{ Al atoms}$

d. $5040 \text{ amu} \times \dfrac{1 \text{ H atom}}{1.008 \text{ amu}} = 5.0 \times 10^3 \text{ H atoms}$

e. $367{,}495.15 \text{ amu} \times \dfrac{1 \text{ Na atom}}{22.99 \text{ amu}} = 1.599 \times 10^4 \text{ Na atoms}$

8. $1.98 \times 10^{13} \text{ amu} \times \dfrac{1 \text{ Na atom}}{22.99 \text{ amu}} = 8.61 \times 10^{11} \text{ Na atoms}$

$3.01 \times 10^{23} \text{ Na atoms} \times \dfrac{22.99 \text{ amu}}{1 \text{ Na atom}} = 6.92 \times 10^{24} \text{ amu}$

## 9.3 The Mole

10. $6.022 \times 10^{23}$ (Avogadro's number)

12. The ratio of the atomic mass of C to the atomic mass of H is (12.01 amu/1.008 amu), and the mass of carbon is given by

$$1.008 \text{ g} \times \frac{12.01 \text{ amu}}{1.008 \text{ amu}} = 12.01 \text{ g C}$$

14. The ratio of the atomic mass of N to the atomic mass of O is (14.01 amu/16.00 amu) and the mass of nitrogen is given by

$$48 \text{ g} \times \frac{14.01 \text{ amu}}{16.00 \text{ amu}} = 42 \text{ g N}$$

16. 1 mol O = 16.00 g O = $6.02 \times 10^{23}$ O atoms

$$1 \text{ O atom} \times \frac{16.00 \text{ g O}}{6.022 \times 10^{23} \text{ O atoms}} = 2.66 \times 10^{-23} \text{ g}$$

18. $$0.50 \text{ mol O atoms} \times \frac{16.00 \text{ g}}{1 \text{ mol}} = 8.0 \text{ g O}$$

$$4 \text{ mol H atoms} \times \frac{1.008 \text{ g}}{1 \text{ mol}} = 4 \text{ g H}$$

Half a mole of O atoms weighs more than 4 moles of H atoms.

20. a. 1.5 mg = 0.0015 g

$$0.0015 \text{ g Cr} \times \frac{1 \text{ mol}}{52.00 \text{ g}} = 2.9 \times 10^{-5} \text{ mol Cr}$$

b. $$2.0 \times 10^{-3} \text{ g Sr} \times \frac{1 \text{ mol}}{87.62 \text{ g}} = 2.3 \times 10^{-5} \text{ mol Sr}$$

c. $$4.84 \times 10^{4} \text{ g B} \times \frac{1 \text{ mol}}{10.81 \text{ g}} = 4.48 \times 10^{3} \text{ mol B}$$

d. $3.6 \times 10^{-6}\ \mu\text{g} = 3.6 \times 10^{-12}$ g

$$3.6 \times 10^{-12} \text{ g Cf} \times \frac{1 \text{ mol}}{251 \text{ g}} = 1.4 \times 10^{-14} \text{ mol Cf}$$

e. $$2000 \text{ lb} \times \frac{454 \text{ g}}{1 \text{ lb}} = 9.1 \times 10^{5} \text{ g}$$

$$9.1 \times 10^5 \text{ g Fe} \times \frac{1 \text{ mol}}{55.85 \text{ g}} = 1.6 \times 10^4 \text{ mol Fe}$$

f. $$20.4 \text{ g Ba} \times \frac{1 \text{ mol}}{137.3 \text{ g}} = 0.149 \text{ mol Ba}$$

g. $$62.8 \text{ g Co} \times \frac{1 \text{ mol}}{58.93 \text{ g}} = 1.07 \text{ mol Co}$$

22. a. $$5.0 \text{ mol K} \times \frac{39.10 \text{ g}}{1 \text{ mol}} = 195 \text{ g} = 2.0 \times 10^2 \text{ g K}$$

b. $$0.000305 \text{ mol Hg} \times \frac{200.6 \text{ g}}{1 \text{ mol}} = 0.0612 \text{ g Hg}$$

c. $$2.31 \times 10^{-5} \text{ mol Mn} \times \frac{54.94 \text{ g}}{1 \text{ mol}} = 1.27 \times 10^{-3} \text{ g Mn}$$

d. $$10.5 \text{ mol P} \times \frac{30.97 \text{ g}}{1 \text{ mol}} = 325 \text{ g P}$$

e. $$4.9 \times 10^4 \text{ mol Fe} \times \frac{55.85 \text{ g}}{1 \text{ mol}} = 2.7 \times 10^6 \text{ g Fe}$$

f. $$125 \text{ mol Li} \times \frac{6.941 \text{ g}}{1 \text{ mol}} = 868 \text{ g Li}$$

g. $$0.01205 \text{ mol F} \times \frac{19.00 \text{ g}}{1 \text{ mol}} = 0.2290 \text{ g F}$$

24. a. $$2.89 \text{ g Au} \times \frac{6.022 \times 10^{23} \text{ Au atoms}}{197.0 \text{ g Au}} = 8.83 \times 10^{21} \text{ Au atoms}$$

b. $$0.000259 \text{ mol Pt} \times \frac{6.022 \times 10^{23} \text{ Pt atoms}}{1 \text{ mol}} = 1.56 \times 10^{20} \text{ Pt atoms}$$

c. $$0.000259 \text{ g Pt} \times \frac{6.022 \times 10^{23} \text{ Pt atoms}}{195.1 \text{ g Pt}} = 7.99 \times 10^{17} \text{ Pt atoms}$$

d. $$2.0 \text{ lb} \times \frac{454 \text{ g}}{1 \text{ lb}} = 908 \text{ g}$$

$$908 \text{ g Mg} \times \frac{6.022 \times 10^{23} \text{ Mg atoms}}{24.31 \text{ g Mg}} = 2.3 \times 10^{25} \text{ Mg atoms}$$

e. $$1.90 \text{ mL} \times \frac{13.6 \text{ g}}{1 \text{ mL}} = 25.8 \text{ g Hg}$$

$$25.8 \text{ g Hg} \times \frac{6.022 \times 10^{23} \text{ Hg atoms}}{200.6 \text{ g Hg}} = 7.75 \times 10^{22} \text{ Hg atoms}$$

f. $4.30 \text{ mol W} \times \dfrac{6.022 \times 10^{23} \text{ W atoms}}{1 \text{ mol}} = 2.59 \times 10^{24} \text{ W atoms}$

g. $4.30 \text{ g W} \times \dfrac{6.022 \times 10^{23} \text{ W atoms}}{183.9 \text{ g W}} = 1.41 \times 10^{22} \text{ W atoms}$

## 9.4 Molar Mass

Note: The values of molar masses are given here to four significant figures, but sometimes five are used.

26. adding together

28. a.

| | | |
|---|---|---|
| mass of 3 mol C = | 3(16.00 g) = | 48.00 g |
| mass of 8 mol H = | 8(1.008 g) = | 8.064 g |
| molar mass of $C_3H_8$ = | | 44.09 g |

b.

| | | |
|---|---|---|
| mass of 2 mol Fe = | 2(55.85 g) = | 111.7 g |
| mass of 3 mol S = | 3(32.06 g) = | 96.18 g |
| mass of 12 mol O = | 12(16.00 g) = | 192.0 g |
| molar mass of $Fe_2(SO_4)_3$ = | | 399.9 g |

c.

| | | |
|---|---|---|
| mass of 2 mol N = | 2(14.01 g) = | 28.02 g |
| mass of 5 mol O = | 5(16.00 g) = | 80.00 g |
| molar mass of $N_2O_5$ = | | 108.02 g = 108.0 g |

d.

| | | |
|---|---|---|
| mass of 12 mol C = | 12(12.01 g) = | 144.1 g |
| mass of 22 mol H = | 22(1.008 g) = | 22.18 g |
| mass of 11 mol O = | 11(16.00 g) = | 176.0 g |
| molar mass of $C_{12}H_{22}O_{11}$ = | | 342.3 g |

e.

| | | |
|---|---|---|
| mass of 2 mol N = | 2(14.01 g) = | 28.02 g |
| mass of 8 mol H = | 8(1.008 g) = | 8.064 g |
| mass of 1 mol C = | 12.01 g = | 12.01 g |
| mass of 3 mol O = | 3(16.00 g) = | 48.00 g |
| molar mass of $(NH_4)_2CO_3$ = | | 96.09 g |

f.

| | | |
|---|---|---|
| mass of 1 mol Ba = | 137.3 g = | 137.3 g |
| mass of 2 mol Cl = | 2(35.45 g) = | 70.90 g |
| molar mass of $BaCl_2$ = | | 208.2 g |

30. a.

mass of 6 mol C = 6(12.01 g) = 72.06 g

mass of 10 mol H = 10(1.008 g) = 10.08 g

mass of 4 mol O = 4(16.00 g) = 64.00 g

---

molar mass of $C_6H_{10}O_4$ = 146.14 g = 146.1 g

b.

mass of 8 mol C = 8(12.01 g) = 96.08 g

mass of 10 mol H = 10(1.008 g) = 10.08 g

mass of 4 mol N = 4(14.01 g) = 56.04 g

mass of 2 mol O = 1(16.00 g) = 32.00 g

---

molar mass of $C_8H_{10}N_4O_2$ = 194.20 g = 194.2 g

c.

mass of 20 mol C = 20(12.01 g) = 240.2 g

mass of 42 mol H = 42(1.008 g) = 42.34 g

---

molar mass of $C_{20}H_{42}$ = 282.5 g

d.

mass of 6 mol C = 6(12.01 g) = 72.06 g

mass of 12 mol H = 12(1.008 g) = 12.10 g

mass of 1 mol O = 16.00 g = 16.00 g

---

molar mass of $C_6H_{11}OH$ = 100.16 g = 100.2 g

e.

mass of 4 mol C = 4(12.01 g) = 48.04 g

mass of 6 mol H = 6(1.008 g) = 6.048 g

mass of 2 mol O = 2(16.00 g) = 32.00 g

---

molar mass of $C_4H_6O_2$ = 86.09 g

f.

mass of 6 mol C = 6(12.01 g) = 72.06 g

mass of 12 mol H = 12(1.008 g) = 12.10 g

mass of 6 mol O = 6(16.00 g) = 96.00 g

---

molar mass of $C_6H_{12}O_6$ = 180.16 g

32. a.

molar mass of $(NH_4)_2S$ = 68.1 g

$$21.2 \text{ g} \times \frac{1 \text{ mol}}{68.1 \text{ g}} = 0.311 \text{ mol } (NH_4)_2S$$

b.

molar mass of $Ca(NO_3)_2$ = 164.1 g

$$44.3 \text{ g} \times \frac{1 \text{ mol}}{164.1 \text{ g}} = 0.270 \text{ mol } Ca(NO_3)_2$$

c. molar mass of $Cl_2O$ = 86.9 g

$$4.35 \text{ g} \times \frac{1 \text{ mol}}{86.9 \text{ g}} = 0.0501 \text{ mol } Cl_2O$$

d. 1.0 lb = 454 g

molar mass of $FeCl_3$ = 162.2

$$454 \text{ g} \times \frac{1 \text{ mol}}{162.2 \text{ g}} = 2.8 \text{ mol } FeCl_3$$

e. 1.0 kg = $1.0 \times 10^3$ g

molar mass of $FeCl_3$ = 162.2 g

$$1.0 \times 10^3 \text{ g} \times \frac{1 \text{ mol}}{162.2 \text{ g}} = 6.2 \text{ mol } FeCl_3$$

34. a. molar mass of KCN = 65.1 g

$$0.00056 \text{ g} \times \frac{1 \text{ mol}}{65.1 \text{ g}} = 8.6 \times 10^{-6} \text{ mol KCN}$$

b. 1.90 mg = 0.00190 g

molar mass of $C_{11}H_{17}O_3N$ = 211.3 g

$$0.00190 \text{ g} \times \frac{1 \text{ mol}}{211.3 \text{ g}} = 9.00 \times 10^{-6} \text{ mol } C_{11}H_{17}O_3N$$

c. molar mass of $Ca_3(PO_4)_2$ = 310.2 g

$$23.09 \text{ g} \times \frac{1 \text{ mol}}{310.2 \text{ g}} = 0.07444 \text{ mol } Ca_3(PO_4)_2$$

d. $2.6 \times 10^{-3}$ mg = $2.6 \times 10^{-6}$ g

molar mass of $Rb_2O$ = 186.9 g

$$2.6 \times 10^{-6} \text{ g} \times \frac{1 \text{ mol}}{186.9 \text{ g}} = 1.4 \times 10^{-8} \text{ mol } Rb_2O$$

e. 6.0 $\mu$g = $6.0 \times 10^{-6}$ g

molar mass of $NH_4NO_3$ = 80.1 g

$$6.0 \times 10^{-6} \text{ g} \times \frac{1 \text{ mol}}{80.1 \text{ g}} = 7.5 \times 10^{-8} \text{ mol } NH_4NO_3$$

36. a. molar mass of $CuSO_4$ = 159.6 g

$$2.6 \times 10^{-2} \text{ mol} \times \frac{159.6 \text{ g}}{1 \text{ mol}} = 4.1 \text{ g } CuSO_4$$

b. molar mass of $C_2F_4$ = 100.0 g

$$3.05 \times 10^{3} \text{ mol} \times \frac{100.0 \text{ g}}{1 \text{ mol}} = 3.05 \times 10^{5} \text{ g } C_2F_4$$

c. 7.83 millimol = 0.00783 mol

molar mass of $C_5H_8$ = 68.1 g

$$0.00783 \text{ mol} \times \frac{68.1 \text{ g}}{1 \text{ mol}} = 0.533 \text{ g } C_5H_8$$

d. molar mass of $BiCl_3$ = 315.3 g

$$6.30 \text{ mol} \times \frac{315.3 \text{ g}}{1 \text{ mol}} = 1.99 \times 10^{3} \text{ g } BiCl_3$$

e. molar mass of $C_{12}H_{22}O_{11}$ = 342.3 g

$$12.2 \text{ mol} \times \frac{342.3 \text{ g}}{1 \text{ mol}} = 4.18 \times 10^{3} \text{ g } C_{12}H_{22}O_{11}$$

38. a. molar mass of $(NH_4)_2CO_3$ = 96.1 g

$$3.09 \text{ mol} \times \frac{96.1 \text{ g}}{1 \text{ mol}} = 297 \text{ g } (NH_4)_2CO_3$$

b. molar mass of $NaHCO_3$ = 84.0 g

$$4.01 \times 10^{-6} \text{ mol} \times \frac{84.0 \text{ g}}{1 \text{ mol}} = 3.37 \times 10^{-4} \text{ g } NaHCO_3$$

c. molar mass of $CO_2$ = 44.01 g

$$88.02 \text{ mol} \times \frac{44.01 \text{ g}}{1 \text{ mol}} = 3874 \text{ g } CO_2$$

d. 1.29 millimol = 0.00129 mol

molar mass of $AgNO_3$ = 169.9 g

$$0.00129 \text{ mol} \times \frac{169.9 \text{ g}}{1 \text{ mol}} = 0.219 \text{ g } AgNO_3$$

e. molar mass of $CrCl_3$ = 158.4 g

$$0.0024 \text{ mol} \times \frac{158.4 \text{ g}}{1 \text{ mol}} = 0.38 \text{ g } CrCl_3$$

40. a. molar mass of $C_6H_{12}O_6$ = 180.2 g

$$3.45 \text{ g} \times \frac{6.022 \times 10^{23} \text{ molecules}}{180.2 \text{ g}} = 1.15 \times 10^{22} \text{ molecules } C_6H_{12}O_6$$

b.

$$3.45 \text{ mol} \times \frac{6.022 \times 10^{23} \text{ molecules}}{1 \text{ mol}} = 2.08 \times 10^{24} \text{ molecules } C_6H_{12}O_6$$

c. molar mass of $ICl_5$ = 304.2 g

$$25.0 \text{ g} \times \frac{6.022 \times 10^{23} \text{ molecules}}{304.2 \text{ g}} = 4.95 \times 10^{22} \text{ molecules } ICl_5$$

d. molar mass of $B_2H_6$ = 27.67 g

$$1.00 \text{ g} \times \frac{6.022 \times 10^{23} \text{ molecules}}{27.67 \text{ g}} = 2.18 \times 10^{22} \text{ molecules } B_2H_6$$

e. 1.05 millimol = 0.00105 mol

$$0.00105 \text{ mol} \times \frac{6.022 \times 10^{23} \text{ molecules}}{1 \text{ mol}} = 6.32 \times 10^{20} \text{ molecules}$$

42. a. molar mass of $C_6H_6O$ = 94.11 g

$$1.002 \text{ g} \times \frac{1 \text{ mol}}{94.11 \text{ g}} = 0.01065 \text{ mol } C_6H_6O$$

$$0.01065 \text{ mol } C_6H_6O \times \frac{1 \text{ mol O}}{1 \text{ mol } C_6H_6O} = 0.01065 \text{ mol O}$$

b. molar mass of $Na_2O_2$ = 77.98 g

$$4.901 \text{ g} \times \frac{1 \text{ mol}}{77.98 \text{ g}} = 0.06285 \text{ mol } Na_2O_2$$

$$0.06285 \text{ mol } Na_2O_2 \times \frac{2 \text{ mol O}}{1 \text{ mol } Na_2O_2} = 0.1257 \text{ mol O}$$

c. 124.5 mg = 0.1245 g

molar mass of $C_7H_7O_2N$ = 137.1 g

$$0.1245 \text{ g} \times \frac{1 \text{ mol}}{137.1 \text{ g}} = 9.081 \times 10^{-4} \text{ mol } C_7H_7O_2N$$

$$9.081 \times 10^{-4} \text{ mol } C_7H_7O_2N \times \frac{2 \text{ mol O}}{1 \text{ mol } C_7H_7O_2N} = 1.816 \times 10^{-3} \text{ mol O}$$

d. molar mass of $Al(NO_3)_3$ = 213.0 g

$$9.821 \text{ g} \times \frac{1 \text{ mol}}{213.0 \text{ g}} = 0.04611 \text{ mol } Al(NO_3)_3$$

$$0.04611 \text{ mol } Al(NO_3)_3 \times \frac{9 \text{ mol O}}{1 \text{ mol } Al(NO_3)_3} = 0.4150 \text{ mol O}$$

## 9.5 Percent Composition of Compounds

Note: The percentages are usually given to four significant figures here. They are rounded to three significant figures in the text.

44. less

46. a.

| | | |
|---|---|---|
| mass of Ca present = | 3(40.08 g) = | 120.24 g |
| mass of P present = | 2(30.97 g) = | 61.94 g |
| mass of O present = | 8(16.00 g) = | 128.00 g |
| molar mass of $Ca_3(PO_4)_2$ = | | 310.18 g |

$$\% \text{ Ca} = \frac{120.24 \text{ g Ca}}{310.18 \text{ g}} \times 100 = 38.76\% \text{ Ca}$$

$$\% \text{ P} = \frac{61.94 \text{ g P}}{310.18 \text{ g}} \times 100 = 19.97\% \text{ P}$$

$$\% \text{ O} = \frac{128.00 \text{ g O}}{310.18 \text{ g}} \times 100 = 41.27\% \text{ O}$$

b.

| | | |
|---|---|---|
| mass of Cd present = | 112.4 g = | 112.4 g |
| mass of S present = | 32.06 g = | 32.06 g |
| mass of O present = | 4(16.00 g) = | 64.00 g |
| molar mass of $CdSO_4$ = | | 208.5 g |

$$\% \text{ Cd} = \frac{112.4 \text{ g Cd}}{208.5 \text{ g}} \times 100 = 53.91\% \text{ Cd}$$

$$\% \text{ S} = \frac{32.06 \text{ g S}}{208.5 \text{ g}} \times 100 = 15.38\% \text{ S}$$

$$\% \text{ O} = \frac{64.00 \text{ g O}}{208.5 \text{ g}} \times 100 = 30.70\% \text{ O}$$

c.

mass of Fe present = $2(55.85\text{ g}) = 111.7\text{ g}$

mass of S present = $3(32.06\text{ g}) = 96.18\text{ g}$

mass of O present = $12(16.00\text{ g}) = 192.0\text{ g}$

molar mass of $Fe_2(SO_4)_3$ = $399.9\text{ g}$

$$\%\ Fe = \frac{111.7\text{ g Fe}}{399.9\text{ g}} \times 100 = 27.93\%\ Fe$$

$$\%\ S = \frac{96.18\text{ g S}}{399.9\text{ g}} \times 100 = 24.05\%\ S$$

$$\%\ O = \frac{192.0\text{ g O}}{399.9\text{ g}} \times 100 = 48.01\%\ O$$

d.

mass of Mn present = $54.94\text{ g} = 54.94\text{ g}$

mass of Cl present = $2(35.45\text{ g}) = 70.90\text{ g}$

molar mass of $MnCl_2$ = $125.84\text{ g}$

$$\%\ Mn = \frac{54.94\text{ g Mn}}{125.84\text{ g}} \times 100 = 43.66\%\ Mn$$

$$\%\ Cl = \frac{70.90\text{ g Cl}}{125.84\text{ g}} \times 100 = 56.34\%\ Cl$$

e.

mass of N present = $2(14.01\text{ g}) = 28.02\text{ g}$

mass of H present = $8(1.008\text{ g}) = 8.064\text{ g}$

mass of C present = $12.01\text{ g} = 12.01\text{ g}$

mass of O present = $3(16.00\text{ g}) = 48.00\text{ g}$

molar mass of $(NH_4)_2CO_3$ = $96.09\text{ g}$

$$\%\ N = \frac{28.02\text{ g N}}{96.09\text{ g}} \times 100 = 29.16\%\ N$$

$$\%\ H = \frac{8.064\text{ g H}}{96.09\text{ g}} \times 100 = 8.392\%\ H$$

$$\%\ C = \frac{12.01\text{ g C}}{96.09\text{ g}} \times 100 = 12.50\%\ C$$

$$\%\ O = \frac{48.00\text{ g O}}{96.09\text{ g}} \times 100 = 49.95\%\ O$$

f.

| | | |
|---|---|---|
| mass of Na present = | 22.99 g = | 22.99 g |
| mass of H present = | 1.008 g = | 1.008 g |
| mass of C present = | 12.01 g = | 12.01 g |
| mass of O present = | 3(16.00 g) = | 48.00 g |
| molar mass of $NaHCO_3$ = 84.01 g | | |

$$\% \text{ Na} = \frac{22.99 \text{ g Na}}{84.01 \text{ g}} \times 100 = 27.37\% \text{ Na}$$

$$\% \text{ H} = \frac{1.008 \text{ g H}}{84.01 \text{ g}} \times 100 = 1.200\% \text{ H}$$

$$\% \text{ C} = \frac{12.01 \text{ g C}}{84.01 \text{ g}} \times 100 = 14.30\% \text{ C}$$

$$\% \text{ O} = \frac{48.00 \text{ g O}}{84.01 \text{ g}} \times 100 = 57.14\% \text{ O}$$

g.

| | | |
|---|---|---|
| mass of C present = | 12.01 g = | 12.01 g |
| mass of O present = | 2(16.00 g) = | 32.00 g |
| molar mass of $CO_2$ = | | 44.01 g |

$$\% \text{ C} = \frac{12.01 \text{ g C}}{44.01 \text{ g}} \times 100 = 27.29\% \text{ C}$$

$$\% \text{ O} = \frac{32.00 \text{ g O}}{44.01 \text{ g}} \times 100 = 72.71\% \text{ O}$$

h.

| | | |
|---|---|---|
| mass of Ag present = | 107.9 g = | 107.9 g |
| mass of N present = | 14.01 g = | 14.01 g |
| mass of O present = | 3(16.00 g) = | 48.00 g |
| molar mass of $AgNO_3$ = | | 169.9 g |

$$\% \text{ Ag} = \frac{107.9 \text{ g Ag}}{169.9 \text{ g}} \times 100 = 63.51\% \text{ Ag}$$

$$\% \text{ N} = \frac{14.01 \text{ g N}}{169.9 \text{ g}} \times 100 = 8.246\% \text{ N}$$

$$\% \text{ O} = \frac{48.00 \text{ g O}}{169.9 \text{ g}} \times 100 = 28.25\% \text{ O}$$

48. a. molar mass of $C_3H_8$ = 44.09 g

$$\% \text{ C} = \frac{36.03 \text{ g C}}{44.09 \text{ g}} \times 100 = 81.72\% \text{ C}$$

b. molar mass of $KClO_3$ = 122.6 g

$$\% K = \frac{39.10 \text{ g K}}{122.6 \text{ g}} \times 100 = 31.89\% \text{ K}$$

c. molar mass of $N_2O_5$ = 108.02 g

$$\% N = \frac{28.02 \text{ g N}}{108.02 \text{ g}} \times 100 = 25.94\% \text{ N}$$

d. molar mass of $C_{12}H_{22}O_{11}$ = 342.3 g

$$\% C = \frac{144.1 \text{ g C}}{342.3 \text{ g}} \times 100 = 42.10\% \text{ C}$$

e. molar mass of $NH_4F$ = 37.09 g

$$\% N = \frac{14.01 \text{ g N}}{37.09 \text{ g}} \times 100 = 37.82\% \text{ N}$$

f. molar mass of $NaClO_4$ = 122.44 g

$$\% Na = \frac{22.99 \text{ g Na}}{122.44 \text{ g}} \times 100 = 18.78\% \text{ Na}$$

g. molar mass of $(NH_4)_2CO_3$ = 96.09 g

$$\% N = \frac{28.02 \text{ g N}}{96.09 \text{ g}} \times 100 = 29.16\% \text{ N}$$

h. molar mass of $BaCl_2$ = 208.2 g

$$\% Ba = \frac{137.3 \text{ g Ba}}{208.2 \text{ g}} \times 100 = 65.95\% \text{ Ba}$$

50. a. $$\% Fe = \frac{55.85 \text{ g Fe}}{151.91 \text{ g}} \times 100 = 36.77\% \text{ Fe}$$

b. $$\% Ag = \frac{215.8 \text{ g Ag}}{231.8 \text{ g}} \times 100 = 93.10\% \text{ Ag}$$

c. $$\% Sr = \frac{87.62 \text{ g Sr}}{158.5 \text{ g}} \times 100 = 55.28\% \text{ Sr}$$

d. $$\% C = \frac{48.04 \text{ g C}}{86.08 \text{ g}} \times 100 = 55.81\% \text{ C}$$

e. $$\% C = \frac{12.01 \text{ g C}}{34.04 \text{ g}} \times 100 = 37.48\% \text{ C}$$

f. $\% \text{ Al} = \dfrac{53.96 \text{ g Al}}{101.96 \text{ g}} \times 100 = 52.92\% \text{ Al}$

g. $\% \text{ K} = \dfrac{39.10 \text{ g K}}{106.55 \text{ g}} \times 100 = 36.70\% \text{ K}$

h. $\% \text{ K} = \dfrac{39.10 \text{ g K}}{74.55 \text{ g}} \times 100 = 52.45\% \text{ K}$

52. a. molar mass of $Ca_3(PO_4)_2$ = 310.18 g

molar mass of $PO_4^{3-}$ ion = 94.97 g

$\% \ PO_4^{3-} = \dfrac{189.9 \text{ g } PO_4^{3-}}{310.18 \text{ g}} \times 100 = 61.22\% \ PO_4^{3-}$

b. molar mass of $CdSO_4$ = 208.5 g

molar mass of $SO_4^{2-}$ ion = 96.06 g

$\% \ SO_4^{2-} = \dfrac{96.06 \text{ g } SO_4^{2-}}{208.5 \text{ g}} \times 100 = 46.07\% \ SO_4^{2-}$

c. molar mass of $Fe_2(SO_4)_3$ = 399.9 g

molar mass of $SO_4^{2-}$ ion = 96.06 g

$\% \ SO_4^{2-} = \dfrac{288.2 \text{ g } SO_4^{2-}}{399.9 \text{ g}} \times 100 = 72.07\% \ SO_4^{2-}$

d. molar mass of $MnCl_2$ = 125.84 g

molar mass of $Cl^-$ ion = 35.45 g

$\% \ Cl^- = \dfrac{70.90 \text{ g } Cl^-}{125.84 \text{ g}} \times 100 = 56.34\% \ Cl^-$

## 9.6 Formulas of Compounds

54. The empirical formula represents the smallest whole number ratio of the elements present in a compound. The molecular formula indicates the actual number of atoms of each element found in a molecule of the substance.

56. a yes (each of these has empirical formula CH)
b. no (the number of hydrogen atoms is wrong)
c. yes (both have empirical formula $NO_2$)
d. no (the number of hydrogen atoms is wrong)

## 9.7 Calculation of Empirical Formulas

58. $0.7238 \text{ g C} \times \dfrac{1 \text{ mol C}}{12.01 \text{ g C}} = 0.06027 \text{ mol C}$

$$0.07088 \text{ g H} \times \frac{1 \text{ mol H}}{1.008 \text{ g H}} = 0.07032 \text{ mol H}$$

$$0.1407 \text{ g N} \times \frac{1 \text{ mol N}}{14.01 \text{ g N}} = 0.01004 \text{ mol N}$$

$$0.3214 \text{ g O} \times \frac{1 \text{ mol O}}{16.00 \text{ g O}} = 0.02009 \text{ mol O}$$

Dividing each number of moles by the smallest number of moles (0.01004 mol N) gives

$$\frac{0.06027 \text{ mol C}}{0.01004} = 6.002 \text{ mol C}$$

$$\frac{0.07032 \text{ mol H}}{0.01004} = 7.004 \text{ mol H}$$

$$\frac{0.01004 \text{ mol N}}{0.01004} = 1.000 \text{ mol N}$$

$$\frac{0.02009 \text{ mol O}}{0.01004} = 2.001 \text{ mol O}$$

The empirical formula is $C_6H_7NO_2$

60. $$0.02990 \text{ g C} \times \frac{1 \text{ mol C}}{12.01 \text{ g C}} = 0.02490 \text{ mol C}$$

$$0.05849 \text{ g H} \times \frac{1 \text{ mol H}}{1.008 \text{ g H}} = 0.05803 \text{ mol H}$$

$$0.2318 \text{ g N} \times \frac{1 \text{ mol N}}{14.01 \text{ g N}} = 0.01655 \text{ mol N}$$

$$0.1328 \text{ g O} \times \frac{1 \text{ mol O}}{16.00 \text{ g O}} = 0.008300 \text{ mol O}$$

Dividing each number of moles by the smallest number of moles (0.008300 mol O) gives

$$\frac{0.02490 \text{ mol C}}{0.008300} = 3.000 \text{ mol C} \qquad \frac{0.05803 \text{ mol H}}{0.008300} = 6.992 \text{ mol H}$$

$$\frac{0.01655 \text{ mol N}}{0.008300} = 1.993 \text{ mol N} \qquad \frac{0.008300 \text{ mol O}}{0.008300} = 1.000 \text{ mol O}$$

The empirical formula is $C_3H_7N_2O$.

62. $2.004 \text{ g Ca} \times \dfrac{1 \text{ mol Ca}}{40.08 \text{ g Ca}} = 0.05000 \text{ mol Ca}$

$0.4670 \text{ g N} \times \dfrac{1 \text{ mol Cl}}{14.01 \text{ g N}} = 0.03333 \text{ mol N}$

Dividing each number of moles by the smaller number of moles (0.03333 mol N) gives

$\dfrac{0.05000 \text{ mol Ca}}{0.03333} = 1.500 \text{ mol Ca}$ $\dfrac{0.03333 \text{ mol N}}{0.03333} = 1.000 \text{ mol N}$

Multiplying these relative numbers of moles by 2 to give whole numbers gives the empirical formula as $Ca_3N_2$.

64. Mass of oxygen in compound = 4.33 g − 4.01 g = 0.32 g O

$4.01 \text{ g Hg} \times \dfrac{1 \text{ mol Hg}}{200.6 \text{ g Hg}} = 0.0200 \text{ mol Hg}$

$0.32 \text{ g O} \times \dfrac{1 \text{ mol O}}{16.00 \text{ g O}} = 0.020 \text{ mol O}$

Since the numbers of moles are equal, the empirical formula is HgO.

66. Mass of chlorine in compound = 3.045 g − 1.00 g = 2.04 g Cl

$1.00 \text{ g Cr} \times \dfrac{1 \text{ mol Cr}}{52.00 \text{ g Cr}} = 0.0192 \text{ mol Cr}$

$2.04 \text{ g Cl} \times \dfrac{1 \text{ mol Cl}}{35.45 \text{ g Cl}} = 0.0577 \text{ mol Cl}$

Dividing each number of moles by the smaller number of moles (0.0192 mol Cr) gives

$\dfrac{0.0192 \text{ mol Cr}}{0.0192} = 1.00 \text{ mol Cr}$ $\dfrac{0.0577 \text{ mol Cl}}{0.0192} = 3.01 \text{ mol Cl}$

The empirical formula is $CrCl_3$.

68. Consider 100.0 g of the compound.

$65.95 \text{ g Ba} \times \dfrac{1 \text{ mol Ba}}{137.3 \text{ g Ba}} = 0.4803 \text{ mol Ba}$

$34.05 \text{ g Cl} \times \dfrac{1 \text{ mol Cl}}{35.45 \text{ g Cl}} = 0.9605 \text{ mol Cl}$

Dividing each number of moles by the smaller number of moles (0.4803 mol Ba) gives

$$\frac{0.4803 \text{ mol Ba}}{0.4803} = 1.000 \text{ mol Ba} \qquad \frac{0.9605 \text{ mol Cl}}{0.4803} = 2.000 \text{ mol Cl}$$

The empirical formula is $BaCl_2$.

70. Consider 100.0 g of the compound.

$$59.78 \text{ g Li} \times \frac{1 \text{ mol Li}}{6.941 \text{ g Li}} = 8.612 \text{ mol Li}$$

$$40.22 \text{ g N} \times \frac{1 \text{ mol N}}{14.00 \text{ g N}} = 2.871 \text{ mol N}$$

Dividing each number of moles by the smaller number of moles (2.871 mol N) gives

$$\frac{8.612 \text{ mol Li}}{2.871} = 2.999 \text{ mol Li} \qquad \frac{2.871 \text{ mol N}}{2.871} = 1.000 \text{ mol N}$$

The empirical formula is $Li_3N$.

72. Consider 100.0 g of the compound.

$$15.77 \text{ g Al} \times \frac{1 \text{ mol Al}}{26.98 \text{ g Al}} = 0.5845 \text{ mol Al}$$

$$28.11 \text{ g S} \times \frac{1 \text{ mol S}}{32.06 \text{ g S}} = 0.8768 \text{ mol S}$$

$$56.12 \text{ g O} \times \frac{1 \text{ mol O}}{16.00 \text{ g O}} = 3.508 \text{ mol O}$$

Dividing each number of moles by the smallest number of moles (0.5845 mol Al) gives

$$\frac{0.5845 \text{ mol Al}}{0.5845} = 1.000 \text{ mol Al} \qquad \frac{0.8768 \text{ mol S}}{0.5845} = 1.500 \text{ mol S}$$

$$\frac{3.508 \text{ mol O}}{0.5845} = 6.002 \text{ mol O}$$

Multiplying these relative numbers of moles by 2 to give whole numbers gives the empirical formula as $Al_2S_3O_{12}$.

74. Consider 100.0 g of the compound.

$$21.60 \text{ g Na} \times \frac{1 \text{ mol Na}}{22.98 \text{ g Na}} = 0.9396 \text{ mol Na}$$

$$33.32 \text{ g Cl} \times \frac{1 \text{ mol Cl}}{35.45 \text{ g Cl}} = 0.9399 \text{ mol Cl}$$

$$45.09 \text{ g O} \times \frac{1 \text{ mol O}}{16.00 \text{ g O}} = 2.818 \text{ mol O}$$

Dividing each number of moles by the smallest number of moles (0.9396 mol Na) gives

$$\frac{0.9396 \text{ mol Na}}{0.9396} = 1.000 \text{ mol Na} \qquad \frac{0.9399 \text{ mol Cl}}{0.9396} = 1.000 \text{ mol Cl}$$

$$\frac{2.818 \text{ mol O}}{0.9396} = 2.999 \text{ mol O}$$

The empirical formula is $NaClO_3$.

## 9.8 Calculation of Molecular Formulas

76. If only the empirical formula is known, the molar mass of the substance must be determined before the molecular formula can be calculated.

78. empirical formula mass of $C_2HF_4$ = 101 g

$$n = \frac{\text{molar mass}}{\text{empirical formula mass}} = \frac{200 \text{ g}}{101 \text{ g}} = 2$$

molecular formula is $(C_2HF_4)_2$ or $C_4H_2F_8$.

80. empirical formula mass of $NO_2$ = 46 g

$$n = \frac{\text{molar mass}}{\text{empirical formula mass}} = \frac{92 \text{ g}}{46 \text{ g}} = 2$$

molecular formula is $(NO_2)_2 = N_2O_4$.

82. Consider 100.0 g of the compound.

$$65.45 \text{ g C} \times \frac{1 \text{ mol C}}{12.01 \text{ g C}} = 5.450 \text{ mol C}$$

$$5.492 \text{ g H} \times \frac{1 \text{ mol H}}{1.008 \text{ g H}} = 5.448 \text{ mol H}$$

$$29.06 \text{ g O} \times \frac{1 \text{ mol O}}{16.00 \text{ g O}} = 1.816 \text{ mol O}$$

Dividing each number of moles by the smallest number of moles (1.816 mol O) gives

$$\frac{5.450 \text{ mol C}}{1.816} = 3.000 \text{ mol C} \qquad \frac{5.448 \text{ mol H}}{1.816} = 3.000 \text{ mol H}$$

$$\frac{1.816 \text{ mol O}}{1.816} = 1.000 \text{ mol O}$$

The empirical formula is $C_3H_3O$, and the empirical formula mass is approximately 55 g.

$$n = \frac{\text{molar mass}}{\text{empirical formula mass}} = \frac{110 \text{ g}}{55 \text{ g}} = 2$$

The molecular formula is $(C_3H_3O)_2 = C_6H_6O_2$.

## Additional Problems

84.

| | | |
|---|---|---|
| 5.00 g Al | 0.185 mol | $1.11 \times 10^{23}$ atoms |
| 0.140 g Fe | 0.00250 mol | $1.51 \times 10^{21}$ atoms |
| $2.7 \times 10^{2}$ g Cu | 4.3 mol | $2.6 \times 10^{24}$ atoms |
| 0.00250 g Mg | $1.03 \times 10^{-4}$ mol | $6.19 \times 10^{19}$ atoms |
| 0.062 g Na | $2.7 \times 10^{-3}$ mol | $1.6 \times 10^{21}$ atoms |
| $3.95 \times 10^{-18}$g U | $1.66 \times 10^{-20}$ mol | $1.00 \times 10^{4}$ atoms |

86.

| | | |
|---|---|---|
| mass of 2 mol X = | 2(41.2 g) = | 82.4 g |
| mass of 1 mol Y = | 57.7 g = | 57.7 g |
| mass of 3 mol Z = | 3(63.9 g) = | 191.7 g |
| molar mass of $X_2YZ_3$ = | | 331.8 g |

$$\% \text{ X} = \frac{82.4 \text{ g}}{331.8 \text{ g}} \times 100 = 24.8\% \text{ X}$$

$$\% \text{ Y} = \frac{57.7 \text{ g}}{331.8 \text{ g}} \times 100 = 17.4\% \text{ Y}$$

$$\% \text{ Z} = \frac{191.7 \text{ g}}{331.8 \text{ g}} \times 100 = 57.8\% \text{ Z}$$

If the molecular formula were actually $X_4Y_2Z_6$, the percentage composition would be the same: the *relative* mass of each element present would not change. The molecular formula is always a whole number multiple of the empirical formula.

88. For the first compound (*restricted* amount of oxygen)

$$2.118 \text{ g Cu} \times \frac{1 \text{ mol Cu}}{63.54 \text{ g Cu}} = 0.03333 \text{ mol Cu}$$

$$0.2666 \text{ g O} \times \frac{1 \text{ mol O}}{16.00 \text{ g O}} = 0.01666 \text{ mol O}$$

Since the number of moles of Cu (0.03333 mol) is twice the number of moles of O (0.01666 mol), the empirical formula is $Cu_2O$.

For the second compound (stream of pure oxygen)

$$2.118 \text{ g Cu} \times \frac{1 \text{ mol Cu}}{63.54 \text{ g Cu}} = 0.03333 \text{ mol Cu}$$

$$0.5332 \text{ g O} \times \frac{1 \text{ mol O}}{16.00 \text{ g O}} = 0.03333 \text{ mol O}$$

Since the numbers of moles are the same, the empirical formula is CuO.

90. a. molar mass of $Pb(NO_3)_2$ = 331.2 g

$$2.37 \text{ g} \times \frac{1 \text{ mol}}{331.2 \text{ g}} = 7.15 \times 10^{-3} \text{ mol } Pb(NO_3)_2$$

$$7.15 \times 10^{-3} \text{ mol } Pb(NO_3)_2 \times \frac{1 \text{ mol Pb}}{1 \text{ mol } Pb(NO_3)_2} = 7.15 \times 10^{-3} \text{ mol Pb}$$

$$7.15 \times 10^{-3} \text{ mol Pb} \times \frac{6.022 \times 10^{23} \text{ atoms}}{1 \text{ mol}} = 4.31 \times 10^{21} \text{ Pb atoms}$$

$$7.15 \times 10^{-3} \text{ mol } Pb(NO_3)_2 \times \frac{2 \text{ mol N}}{1 \text{ mol } Pb(NO_3)_2} = 0.0143 \text{ mol N}$$

$$0.0143 \text{ mol N} \times \frac{6.022 \times 10^{23} \text{ atoms}}{1 \text{ mol}} = 8.62 \times 10^{21} \text{ N atoms}$$

$$7.15 \times 10^{-3} \text{ mol } Pb(NO_3)_2 \times \frac{6 \text{ mol O}}{1 \text{ mol } Pb(NO_3)_2} = 0.0429 \text{ mol O}$$

$$0.0429 \text{ mol O} \times \frac{6.022 \times 10^{23} \text{ atoms}}{1 \text{ mol}} = 2.58 \times 10^{22} \text{ O atoms}$$

b. molar mass $O_2$ = 32.00 g

$$22.4 \text{ g } O_2 \times \frac{1 \text{ mol}}{32.00 \text{ g}} = 0.700 \text{ mol } O_2$$

$$0.700 \text{ mol } O_2 \times \frac{2 \text{ mol O atoms}}{1 \text{ mol } O_2} = 1.40 \text{ mol O atoms}$$

$$1.40 \text{ mol O} \times \frac{6.022 \times 10^{23} \text{ atoms}}{1 \text{ mol}} = 8.43 \times 10^{23} \text{ O atoms}$$

c. molar mass $S_8$ = 256.48 g

101.9 mg = 0.1019 g

$$0.1019 \text{ g} \times \frac{1 \text{ mol } S_8}{256.48 \text{ g}} = 3.973 \times 10^{-4} \text{ mol } S_8$$

$$3.973 \times 10^{-4} \text{ mol } S_8 \times \frac{8 \text{ mol S}}{1 \text{ mol } S_8} = 3.178 \times 10^{-3} \text{ mol S}$$

$$3.178 \times 10^{-3} \text{ mol S} \times \frac{6.022 \times 10^{23} \text{ S atoms}}{1 \text{ mol}} = 1.914 \times 10^{21} \text{ S atoms}$$

d. molar mass $UF_6$ = 352.0 g

$43.7\ \mu\text{g} = 43.7 \times 10^{-6}$ g

$$43.7 \times 10^{-6} \text{ g} \times \frac{1 \text{ mol}}{352.0 \text{ g}} = 1.24 \times 10^{-7} \text{ mol } UF_6$$

$$1.24 \times 10^{-7} \text{ mol } UF_6 \times \frac{1 \text{ mol U}}{1 \text{ mol } UF_6} = 1.24 \times 10^{-7} \text{ mol U}$$

$$1.24 \times 10^{-7} \text{ mol U} \times \frac{6.022 \times 10^{23} \text{ U atoms}}{1 \text{ mol}} = 7.47 \times 10^{16} \text{ U atoms}$$

$$1.24 \times 10^{-7} \text{ mol } UF_6 \times \frac{6 \text{ mol F}}{1 \text{ mol } UF_6} = 7.44 \times 10^{-7} \text{ mol F}$$

$$7.44 \times 10^{-7} \text{ mol F} \times \frac{6.022 \times 10^{23} \text{ F atoms}}{1 \text{ mol}} = 4.48 \times 10^{17} \text{ F atoms}$$

92. a. molar mass of $C_3O_2$ = 3(12.01 g) + 2(16.00 g) = 68.03 g

$$\% \text{ C} = \frac{36.03 \text{ g}}{68.03 \text{ g}} = 52.96\% \text{ C}$$

$$7.819 \text{ g } C_3O_2 \times \frac{52.96 \text{ g C}}{100.0 \text{ g } C_3O_2} = 4.141 \text{ g C}$$

$$4.141 \text{ g C} \times \frac{6.022 \times 10^{23} \text{ C atoms}}{12.01 \text{ g C}} = 2.076 \times 10^{23} \text{ C atoms}$$

b. molar mass of CO = 12.01 g + 16.00 g = 28.01 g

$$\% \text{ C} = \frac{12.01 \text{ g}}{28.01 \text{ g}} \times 100 = 42.9\% \text{ C}$$

$$1.53 \times 10^{21} \text{ molecules CO} \times \frac{1 \text{ C atom}}{1 \text{ molecule CO}} = 1.53 \times 10^{21} \text{ C atoms}$$

$$1.53 \times 10^{21} \text{ C atoms} \times \frac{12.01 \text{ g C}}{6.022 \times 10^{23} \text{ C atoms}} = 0.0305 \text{ g C}$$

c. molar mass of $C_6H_6O$ = 6(12.01 g) + 6(1.008 g) + 16.00 g = 94.11 g

$$\% \text{ C} = \frac{72.06 \text{ g}}{94.11 \text{ g}} \times 100 = 76.57\% \text{ C}$$

$$0.200 \text{ mol } C_6H_6O \times \frac{6 \text{ mol C}}{1 \text{ mol } C_6H_6O} = 1.20 \text{ mol C}$$

$$1.20 \text{ mol C} \times \frac{12.01 \text{ g C}}{1 \text{ mol C}} = 14.4 \text{ g C}$$

$$14.4 \text{ g C} \times \frac{6.022 \times 10^{23} \text{ C atoms}}{12.01 \text{ g C}} = 7.22 \times 10^{23} \text{ C atoms}$$

94. $$1.00 \text{ g Ne} \times \frac{207.2 \text{ g Pb}}{20.18 \text{ g Ne}} = 10.3 \text{ g Pb}$$

96. $$5.00 \text{ g Te} \times \frac{200.6 \text{ g Hg}}{127.6 \text{ g Te}} = 7.86 \text{ g Hg}$$

98. 153.8 g $CCl_4$ = $6.022 \times 10^{23}$ molecules $CCL_4$

$$1 \text{ molecule} \times \frac{153.8 \text{ g}}{6.022 \times 10^{23} \text{ molecules}} = 2.554 \times 10^{-22} \text{ g}$$

100. a. molar mass of $C_2H_5O_2N$ =

2(12.01 g) + 5(1.008 g) + 2(16.00 g) + 14.01 g = 75.07 g

$$\text{mass fraction N} = \frac{14.01 \text{ g}}{75.07 \text{ g}}$$

$$5.000 \text{ g} \times \frac{14.01 \text{ g}}{75.07 \text{ g}} = 0.9331 \text{ g N}$$

b. molar mass of $Mg_3N_2$ = 3(24.31 g) + 2(14.01 g) = 100.95 g

$$\text{mass fraction N} = \frac{28.02 \text{ g}}{100.95 \text{ g}}$$

$$5.000 \text{ g} \times \frac{28.02 \text{ g}}{100.95 \text{ g}} = 1.388 \text{ g N}$$

c. molar mass of $Ca(NO_3)_2$ =

40.08 g + 2(14.01 g) + 6(16.00 g) = 164.10 g

$$\text{mass fraction N} = \frac{28.02\text{ g}}{164.10\text{ g}}$$

$$5.000\text{ g} \times \frac{28.02\text{ g}}{164.10\text{ g}} = 0.8537\text{ g N}$$

d. molar mass of $N_2O_4$ = 2(14.01 g) + 4(16.00 g) = 92.02 g

$$\text{mass fraction N} = \frac{28.02\text{ g}}{92.02\text{ g}}$$

$$5.000\text{ g} \times \frac{28.02\text{ g}}{92.02\text{ g}} = 1.522\text{ g N}$$

102. Consider 100.0 g of the compound.

$$16.39\text{ g Mg} \times \frac{1\text{ mol Mg}}{24.31\text{ g Mg}} = 0.6742\text{ mol Mg}$$

$$18.89\text{ g N} \times \frac{1\text{ mol N}}{14.01\text{ g N}} = 1.348\text{ mol N}$$

$$64.72\text{ g O} \times \frac{1\text{ mol O}}{16.00\text{ g O}} = 4.045\text{ mol O}$$

Dividing each number of moles by the smallest number of moles (0.6742 mol Mg) gives

$$\frac{0.6742\text{ mol Mg}}{0.6742} = 1.000\text{ mol Mg} \qquad \frac{1.348\text{ mol N}}{0.6742} = 1.999\text{ mol N}$$

$$\frac{4.045\text{ mol O}}{0.6742} = 5.999\text{ mol O}$$

The empirical formula is $MgN_2O_6$.

## Chapter Ten    Chemical Quantities

### 10.1 Information Given by Chemical Equations

2. The coefficients of the balanced chemical equation for a reaction indicate the *relative numbers of moles* of each reactant that combine during the process, as well as the number of moles of each product formed.

4. Balanced chemical equations tell us in what proportions *on a mole basis* substances combine; since the molar masses of $C(s)$ and $O_2(g)$ are different, 1 g of $O_2$ could not represent the same number of moles as 1 g of C.

6 a. $UO_2(s) + 4HF(aq) \rightarrow UF_4(aq) + 2H_2O(l)$

One molecule of uranium(IV) oxide will combine with four molecules of hydrofluoric acid, producing one uranium(IV) fluoride molecule and two water molecules. One mole of uranium(IV) oxide will combine with four moles of hydrofluoric acid to produce one mole of uranium(IV) fluoride and two moles of water.

b. $2NaC_2H_3O_2(aq) + H_2SO_4(aq) \rightarrow Na_2SO_4(aq) + 2HC_2H_3O_2(aq)$

Two molecules (formula units) of sodium acetate react exactly with one molecule of sulfuric acid, producing one molecule (formula unit) of sodium acetate and two molecules of acetic acid. Two moles of sodium acetate will combine with one mole of sulfuric acid, producing one mole of sodium sulfate and two moles of acetic acid.

c. $Mg(s) + 2HCl(aq) \rightarrow MgCl_2(aq) + H_2(g)$

One magnesium atom will react with two hydrochloric acid molecules (formula units) to produce one molecule (formula unit) of magnesium chloride and one molecule of hydrogen gas. One mole of magnesium will combine with two moles of hydrochloric acid, producing one mole of magnesium chloride and one mole of gaseous hydrogen.

d. $B_2O_3(s) + 3H_2O(l) \rightarrow 2B(OH)_3(s)$

One molecule (formula unit) of diboron trioxide will react exactly with three molecules of water, producing two molecules of boron trihydroxide (boric acid). One mole of diboron trioxide will combine with three moles of water to produce two moles of boron trihydroxide (boric acid).

### 10.2 Mole-Mole Relationships

8. False. For 0.40 mol of $Mg(OH)_2$ to react, 0.80 mol of HCl will be needed. According to the balanced equation, for a given amount of $Mg(OH)_2$, *twice* as many moles of HCl is needed.

10. For $O_2$: $\dfrac{5 \text{ mol } O_2}{1 \text{ mol } C_3H_8}$    For $CO_2$: $\dfrac{3 \text{ mol } CO_2}{1 \text{ mol } C_3H_8}$    For $H_2O$: $\dfrac{4 \text{ mol } H_2O}{1 \text{ mol } C_3H_8}$

12. a. $2H_2O_2(l) \rightarrow 2H_2O(l) + O_2(g)$

$$0.50 \text{ mol } H_2O_2 \times \frac{2 \text{ mol } H_2O}{2 \text{ mol } H_2O_2} = 0.50 \text{ mol } H_2O$$

$$0.50 \text{ mol } H_2O_2 \times \frac{1 \text{ mol } O_2}{2 \text{ mol } H_2O_2} = 0.25 \text{ mol } O_2$$

b. $2KClO_3(s) \rightarrow 2KCl(s) + 3O_2(g)$

$$0.50 \text{ mol } KClO_3 \times \frac{2 \text{ mol } KCl}{2 \text{ mol } KClO_3} = 0.50 \text{ mol } KCl$$

$$0.50 \text{ mol } KClO_3 \times \frac{3 \text{ mol } O_2}{2 \text{ mol } KClO_3} = 0.75 \text{ mol } O_2$$

c. $2Al(s) + 6HCl(aq) \rightarrow 2AlCl_3(aq) + 3H_2(g)$

$$0.50 \text{ mol Al} \times \frac{2 \text{ mol } AlCl_3}{2 \text{ mol Al}} = 0.50 \text{ mol } AlCl_3$$

$$0.50 \text{ mol Al} \times \frac{3 \text{ mol } H_2}{2 \text{ mol Al}} = 0.75 \text{ mol } H_2$$

d. $C_3H_8(g) + 5O_2(g) \rightarrow 3CO_2(g) + 4H_2O(l)$

$$0.50 \text{ mol } C_3H_8 \times \frac{3 \text{ mol } CO_2}{1 \text{ mol } C_3H_8} = 1.5 \text{ mol } CO_2$$

$$0.50 \text{ mol } C_3H_8 \times \frac{4 \text{ mol } H_2O}{1 \text{ mol } C_3H_8} = 2.0 \text{ mol } H_2O$$

14. a. $NH_3(g) + HCl(g) \rightarrow NH_4Cl(s)$

molar mass of $NH_3$ = 17.0 g

$$1.00 \text{ g } NH_3 \times \frac{1 \text{ mol } NH_3}{17.0 \text{ g } NH_3} = 0.0588 \text{ mol } NH_3$$

$$0.0588 \text{ mol } NH_3 \times \frac{1 \text{ mol } NH_4Cl}{1 \text{ mol } NH_3} = 0.0588 \text{ mol } NH_4Cl$$

b. $\mathbf{CaO}(s) + CO_2(g) \rightarrow CaCO_3(s)$

molar mass CaO = 56.1 g

$$1.00 \text{ g CaO} \times \frac{1 \text{ mol CaO}}{56.1 \text{ g CaP}} = 0.0178 \text{ mol CaO}$$

$$0.0178 \text{ mol CaO} \times \frac{1 \text{ mol } CaCO_3}{1 \text{ mol CaO}} = 0.0178 \text{ mol } CaCO_3$$

c. $\mathbf{4Na}(s) + O_2(g) \rightarrow 2Na_2O(s)$

molar mass Na = 23.0 g

$$1.00 \text{ g Na} \times \frac{1 \text{ mol Na}}{23.0 \text{ g Na}} = 0.0435 \text{ mol Na}$$

$$0.0435 \text{ mol Na} \times \frac{2 \text{ mol } Na_2O}{4 \text{ mol Na}} = 0.0218 \text{ mol } Na_2O$$

d. $\mathbf{2P}(s) + 3Cl_2(g) \rightarrow 2PCl_3(l)$

molar mass P = 31.0 g

$$1.00 \text{ g P} \times \frac{1 \text{ mol P}}{31.0 \text{ g P}} = 0.0322 \text{ mol P}$$

$$0.0322 \text{ mol P} \times \frac{2 \text{ mol } PCl_3}{1 \text{ mol P}} = 0.0322 \text{ mol } PCl_3$$

16. Before doing the calculations, the equations must be *balanced*.

a. $N_2(g) + 3H_2(g) \rightarrow 2NH_3(g)$

$$3.125 \text{ mol } N_2 \times \frac{3 \text{ mol } H_2}{1 \text{ mol } N_2} = 9.375 \text{ mol } H_2$$

b. $4Al(s) + 3O_2(g) \rightarrow 2Al_2O_3(s)$

$$3.125 \text{ mol Al} \times \frac{2 \text{ mol } O_2}{4 \text{ mol Al}} = 2.344 \text{ mol } O_2$$

c. $2Ag^+(aq) + CO_3^{2-}(aq) \rightarrow Ag_2CO_3(s)$

$$3.125 \text{ mol } Ag^+ \times \frac{1 \text{ mol } CO_3^{2-}}{2 \text{ mol } Ag^+} = 1.563 \text{ mol } CO_3^{2-}$$

d. $C_5H_{12}(l) + 8O_2(g) \rightarrow 5CO_2(g) + 6H_2O(l)$

$$3.125 \text{ mol } C_5H_{12} \times \frac{8 \text{ mol } O_2}{1 \text{ mol } C_5H_{12}} = 25.00 \text{ mol } O_2$$

## 10.3 Mass Calculations

18. Stoichiometry is the process of using a chemical equation to calculate the relative masses of reactants and products involved in a reaction.

20. a. molar mass Li = 6.941 g

1.5 mg = 0.0015 g

$$0.0015 \text{ g Li} \times \frac{1 \text{ mol Li}}{6.941 \text{ g Li}} = 2.2 \times 10^{-4} \text{ mol Li}$$

b. molar mass $N_2O$ = 44.02 g

$$2.0 \times 10^{-3} \text{ g } N_2O \times \frac{1 \text{ mol } N_2O}{44.02 \text{ g } N_2O} = 4.5 \times 10^{-5} \text{ mol } N_2O$$

c. molar mass $PCl_3$ = 137.3 g

$$4.84 \times 10^{4} \text{ g } PCl_3 \times \frac{1 \text{ mol } PCl_3}{137.3 \text{ g } PCl_3} = 352 \text{ mol } PCl_3$$

d. molar mass $UF_6$ = 352.0 g

$$3.6 \times 10^{-2} \ \mu\text{g } UF_6 \times \frac{1 \text{ mol } UF_6}{252.0 \text{ g } UF_6} = 1.0 \times 10^{-10} \text{ mol } UF_6$$

e. molar mass PbS = 239.3 g

1.0 kg = $1.0 \times 10^3$ g

$$1.0 \times 10^{3} \text{ g} \times \frac{1 \text{ mol PbS}}{239.3 \text{ g PbS}} = 4.2 \text{ mol PbS}$$

f. molar mass $H_2SO_4$ = 98.08 g

$$20.4 \text{ g } H_2SO_4 \times \frac{1 \text{ mol } H_2SO_4}{98.08 \text{ g } H_2SO_4} = 0.208 \text{ mol } H_2SO_4$$

g. molar mass $C_6H_{12}O_6$ = 180.2 g

$$62.8 \text{ g } C_6H_{12}O_6 \times \frac{1 \text{ mol } C_6H_{12}O_6}{180.2 \text{ g } C_6H_{12}O_6} = 0.349 \text{ mol } C_6H_{12}O_6$$

22. a. molar mass $HNO_3$ = 63.0 g

$$5.0 \text{ mol } HNO_3 \times \frac{63.0 \text{ g } HNO_3}{1 \text{ mol } HNO_3} = 3.2 \times 10^2 \text{ g } HNO_3$$

b. molar mass Hg = 200.6 g

$$0.000305 \text{ mol Hg} \times \frac{200.6 \text{ g Hg}}{1 \text{ mol Hg}} = 0.0612 \text{ g Hg}$$

c. molar mass $K_2CrO_4$ = 194.2 g

$$2.31 \times 10^{-5} \text{ mol } K_2CrO_4 \times \frac{194.2 \text{ g } K_2CrO_4}{1 \text{ mol } K_2CrO_4} = 4.49 \times 10^{-3} \text{ g } K_2CrO_4$$

d. molar mass $AlCl_3$ = 133.3 g

$$10.5 \text{ mol } AlCl_3 \times \frac{133.3 \text{ g } AlCl_3}{1 \text{ mol } AlCl_3} = 1.40 \times 10^3 \text{ g } AlCl_3$$

e. molar mass $SF_6$ = 244.8 g

$$4.9 \times 10^4 \text{ mol } SF_6 \times \frac{244.8 \text{ g } SF_6}{1 \text{ mol } SF_6} = 7.2 \times 10^6 \text{ g } SF_6$$

f. molar mass $NH_3$ = 17.0 g

$$125 \text{ mol } NH_3 \times \frac{17.0 \text{ g } NH_3}{1 \text{ mol } NH_3} = 2.13 \times 10^3 \text{ g } NH_3$$

g. molar mass $Na_2O_2$ = 77.98 g

$$0.01205 \text{ mol } Na_2O_2 \times \frac{77.98 \text{ g } Na_2O_2}{1 \text{ mol } Na_2O_2} = 0.9396 \text{ g } Na_2O_2$$

24. Before any calculations are done, the equations must be *balanced*.

a. $2CO(g) + O_2(g) \rightarrow 2CO_2(g)$

Molar mass CO = 28.0 g

$$1.00 \text{ g CO} \times \frac{1 \text{ mol CO}}{28.0 \text{ g CO}} = 0.0357 \text{ mol CO}$$

$$0.0357 \text{ mol CO} \times \frac{1 \text{ mol } O_2}{2 \text{ mol CO}} = 0.0178 \text{ mol } O_2$$

b. $2AgNO_3(aq) + CuSO_4(aq) \rightarrow Ag_2SO_4(s) + Cu(NO_3)_2(aq)$

Molar mass $AgNO_3$ = 169.9 g

$$1.00 \text{ g } AgNO_3 \times \frac{1 \text{ mol } AgNO_3}{169.9 \text{ g } AgNO_3} = 0.00588 \text{ mol } AgNO_3$$

$$0.00588 \text{ mol } AgNO_3 \times \frac{1 \text{ mol } CuSO_4}{2 \text{ mol } AgNO_3} = 0.00294 \text{ mol } CuSO_4$$

c. $2Al(s) + 6HCl(g) \rightarrow 2AlCl_3(s) + 3H_2(g)$

Molar mass Al = 26.98 g

$$1.00 \text{ g Al} \times \frac{1 \text{ mol Al}}{26.98 \text{ g Al}} = 0.0371 \text{ mol Al}$$

$$0.0371 \text{ mol Al} \times \frac{6 \text{ mol HCl}}{2 \text{ mol Al}} = 0.111 \text{ mol HCl}$$

d. $C_3H_8(g) + 5O_2(g) \rightarrow 3CO_2(g) + 4H_2O(g)$

Molar mass $C_3H_8$ = 44.1 g

$$1.00 \text{ g } C_3H_8 \times \frac{1 \text{ mol } C_3H_8}{44.1 \text{ g } C_3H_8} = 0.0227 \text{ mol } C_3H_8$$

$$0.0227 \text{ mol } C_3H_8 \times \frac{5 \text{ mol } O_2}{1 \text{ mol } C_3H_8} = 0.114 \text{ mol } O_2$$

26. Before any calculations are done, the equations must be *balanced*. Since the given and required quantities in this question are given in *milligrams*, it is most convenient to perform the calculations in terms of *millimoles* of the substances involved. One millimole of a substance represents the molar mass of the substance expressed in milligrams.

a. $\mathbf{FeSO_4}(aq) + K_2CO_3(aq) \rightarrow FeCO_3(s) + K_2SO_4(aq)$

millimolar masses: $FeSO_4$, 151.9 mg; $FeCO_3$, 115.9 mg; $K_2SO_4$, 174.3 mg

$$10.0 \text{ mg} \times \frac{1 \text{ mmol } FeSO_4}{151.9 \text{ mg } FeSO_4} = 0.0658 \text{ mmol } FeSO_4$$

$$0.0658 \text{ mmol } FeSO_4 \times \frac{1 \text{ mmol } FeCO_3}{1 \text{ mmol } FeSO_4} \times \frac{115.9 \text{ mg } FeCO_3}{1 \text{ mmol } FeCO_3} = 7.63 \text{ mg } FeCO_3$$

$$0.0658 \text{ mmol } FeSO_4 \times \frac{1 \text{ mmol } K_2SO_4}{1 \text{ mmol } FeSO_4} \times \frac{174.3 \text{ mg } K_2SO_4}{1 \text{ mmol } K_2SO_4} = 11.5 \text{ mg } K_2SO_4$$

b. $\mathbf{4Cr}(s) + 3SnCl_4(l) \rightarrow 4CrCl_3(s) + 3Sn(s)$

millimolar masses: Cr, 52.00 mg; $CrCl_3$, 158.4 mg; Sn, 118.7 mg

$$10.0 \text{ mg Cr} \times \frac{1 \text{ mmol Cr}}{52.00 \text{ mg Cr}} = 0.192 \text{ mmol Cr}$$

$$0.192 \text{ mmol Cr} \times \frac{4 \text{ mmol } CrCl_3}{4 \text{ mmol Cr}} \times \frac{158.4 \text{ mg } CrCl_3}{1 \text{ mmol } CrCl_3} = 30.4 \text{ mg } CrCl_3$$

$$0.192 \text{ mmol Cr} \times \frac{3 \text{ mmol Sn}}{4 \text{ mmol Cr}} \times \frac{118.7 \text{ mg Sn}}{1 \text{ mmol Sn}} = 17.1 \text{ mg Sn}$$

c. $16Fe(s) + \mathbf{3S_8}(s) \rightarrow 8Fe_2S_3(s)$

millimolar masses: $S_8$, 256.5 mg; $Fe_2S_3$, 207.9 mg

$$10.0 \text{ mg } S_8 \times \frac{1 \text{ mmol } S_8}{256.5 \text{ mg } S_8} = 0.0390 \text{ mmol } S_8$$

$$0.0390 \text{ mmol } S_8 \times \frac{8 \text{ mmol } Fe_2S_3}{3 \text{ mmol } S_8} \times \frac{207.9 \text{ mg } Fe_2S_3}{1 \text{ mmol } Fe_2S_3} = 21.6 \text{ mg } Fe_2S_3$$

d. $3Ag(s) + \mathbf{4HNO_3}(aq) \rightarrow 3AgNO_3(aq) + 2H_2O(l) + NO(g)$

millimolar masses: $HNO_3$, 63.0 mg; $AgNO_3$, 169.9 mg
$H_2O$, 18.0 mg; NO, 30.0 mg

$$10.0 \text{ mg } HNO_3 \times \frac{1 \text{ mmol } HNO_3}{63.0 \text{ mg } HNO_3} = 0.159 \text{ mmol } HNO_3$$

$$0.159 \text{ mmol } HNO_3 \times \frac{3 \text{ mmol } AgNO_3}{4 \text{ mmol } HNO_3} \times \frac{169.9 \text{ mg } AgNO_3}{1 \text{ mmol } AgNO_3} = 20.3 \text{ mg } AgNO_3$$

$$0.159 \text{ mmol } HNO_3 \times \frac{2 \text{ mmol } H_2O}{4 \text{ mmol } HNO_3} \times \frac{18.0 \text{ mg } H_2O}{1 \text{ mmol } H_2O} = 1.43 \text{ mg } H_2O$$

$$0.159 \text{ mmol } HNO_3 \times \frac{1 \text{ mmol NO}}{4 \text{ mmol } HNO_3} \times \frac{30.0 \text{ mg NO}}{1 \text{ mmol NO}} = 1.19 \text{ mg NO}$$

28. $2H_2(g) + O_2(g) \rightarrow 2H_2O(g)$

molar masses: $H_2$, 2.016 g; $O_2$, 32.00 g

$$1.00 \text{ g } H_2 \times \frac{1 \text{ mol } H_2}{2.016 \text{ g } H_2} = 0.496 \text{ mol } H_2$$

$$0.496 \text{ mol } H_2 \times \frac{1 \text{ mol } O_2}{2 \text{ mol } H_2} = 0.248 \text{ mol } O_2$$

$$0.248 \text{ mol } O_2 \times \frac{32.00 \text{ g } O_2}{1 \text{ mol } O_2} = 7.94 \text{ g } O_2$$

30. $2SO_2(g) + O_2(g) \rightarrow 2SO_3(g)$

molar masses: $SO_2$, 64.06 g; $SO_3$, 80.06 g

150 kg = $1.5 \times 10^5$ g

$$1.5 \times 10^5 \text{ g } SO_2 \times \frac{1 \text{ mol } SO_2}{64.06 \text{ g } SO_2} = 2.342 \times 10^3 \text{ mol } SO_2$$

$$2.342 \times 10^3 \text{ mol } SO_2 \times \frac{2 \text{ mol } SO_3}{2 \text{ mol } SO_2} = 2.342 \times 10^3 \text{ mol } SO_3$$

$$2.342 \times 10^3 \text{ mol } SO_3 \times \frac{80.06 \text{ g } SO_3}{1 \text{ mol } SO_3} = 1.9 \times 10^5 \text{ g } SO^3 = 1.9 \times 10^2 \text{ kg } SO_3$$

32. $2ZnS(s) + 3O_2(g) \rightarrow 2ZnO(s) + 2SO_2(g)$

molar masses: ZnS, 97.44 g; $SO_2$, 64.06 g

$1.0 \times 10^2$ kg = $1.0 \times 10^5$ g

$$1.0 \times 10^5 \text{ g ZnS} \times \frac{1 \text{ mol ZnS}}{97.44 \text{ g ZnS}} = 1.026 \times 10^3 \text{ mol ZnS}$$

$$1.026 \times 10^3 \text{ mol ZnS} \times \frac{2 \text{ mol } SO_2}{2 \text{ mol ZnS}} = 1.026 \times 10^3 \text{ mol } SO_2$$

$$1.026 \times 10^3 \text{ mol } SO_2 \times \frac{64.06 \text{ g } SO_2}{1 \text{ mol } SO_2} = 6.6 \times 10^4 \text{ g } SO_2 = 66 \text{ kg } SO_2$$

34. $Cl_2(g) + 2NaBr(aq) \rightarrow 2NaCl(aq) + Br_2(l)$

molar masses: $Cl_2$, 70.90 g; $Br_2$, 159.80 g

$$25.0 \text{ g } Cl_2 \times \frac{1 \text{ mol } Cl_2}{70.90 \text{ g } Cl_2} = 0.3526 \text{ mol } Cl_2$$

$$0.3526 \text{ mol } Cl_2 \times \frac{1 \text{ mol } Br_2}{1 \text{ mol } Cl_2} = 0.3526 \text{ mol } Br_2$$

$$0.3526 \text{ mol } Br_2 \times \frac{159.80 \text{ g } Br_2}{1 \text{ mol } Br_2} = 56.3 \text{ g } Br_2$$

36. $Cu(s) + 2AgNO_3(aq) \rightarrow Cu(NO_3)_2(aq) + 2Ag(s)$

millimolar masses: Cu, 63.55 mg; $AgNO_3$, 169.9 mg

$$1.95 \text{ mg } AgNO_3 \times \frac{1 \text{ mmol } AgNO_3}{169.9 \text{ mg } AgNO_3} = 0.01148 \text{ mmol } AgNO_3$$

$$0.01148 \text{ mmol } AgNO_3 \times \frac{1 \text{ mmol Cu}}{2 \text{ mmol } AgNO_3} = 0.005740 \text{ mmol Cu}$$

$$0.005740 \text{ mmol Cu} \times \frac{63.55 \text{ mg Cu}}{1 \text{ mmol Cu}} = 0.365 \text{ mg Cu}$$

38. $Zn(s) + 2HCl(aq) \rightarrow ZnCl_2(aq) + H_2(g)$

molar masses: Zn, 65.38 g; $H_2$, 2.016 g

$$2.50 \text{ g Zn} \times \frac{1 \text{ mol Zn}}{65.38 \text{ g Zn}} = 0.03824 \text{ mol Zn}$$

$$0.03824 \text{ mol Zn} \times \frac{1 \text{ mol } H_2}{1 \text{ mol Zn}} = 0.03824 \text{ mol } H_2$$

$$0.03824 \text{ mol } H_2 \times \frac{2.016 \text{ g } H_2}{1 \text{ mol } H_2} = 0.0771 \text{ g } H_2$$

40. $2C_2H_2(g) + 5O_2(g) \rightarrow 4CO_2(g) + 2H_2O(g)$

molar masses: $C_2H_2$, 26.04 g; $O_2$, 32.00 g

$150 \text{ g} = 1.5 \times 10^2 \text{ g}$

$$1.5 \times 10^2 \text{ g } C_2H_2 \times \frac{1 \text{ mol } C_2H_2}{26.04 \text{ g } C_2H_2} = 5.760 \text{ mol } C_2H_2$$

$$5.760 \text{ mol } C_2H_2 \times \frac{5 \text{ mol } O_2}{1 \text{ mol } C_2H_2} = 14.40 \text{ mol } O_2$$

$$14.40 \text{ mol } O_2 \times \frac{32.00 \text{ g } O_2}{1 \text{ mol } O_2} = 4.6 \times 10^2 \text{ g } O_2$$

## 10.4 Calculations Involving a Limiting Reactant

42. To determine the limiting reactant, first calculate the number of moles of each reactant present. Then determine how these numbers of moles correspond to the stoichiometric ratio indicated by the balanced chemical equation for the reaction. See a detailed Summary of the method on page 306 of the text.

44. A reactant is present *in excess* if there is more of that reactant present than is needed to combine with the limiting reactant for the process. An excess of any reactant does not affect the theoretical yield for a process: the theoretical yield is determined by the limiting reactant.

46. a. $2Na(s) + Br_2(l) \rightarrow 2NaBr(s)$

molar masses: Na, 22.99 g; $Br_2$, 159.8 g; NaBr, 102.9 g

$$5.0 \text{ g Na} \times \frac{1 \text{ mol Na}}{22.99 \text{ g Na}} = 0.2175 \text{ mol Na}$$

$$5.0 \text{ g } Br_2 \times \frac{1 \text{ mol } Br_2}{159.8 \text{ g } Br_2} = 0.03129 \text{ mol } Br_2$$

Intuitively, we would suspect that $Br_2$ is the limiting reactant, since there is much less $Br_2$ than Na on a mole basis. To *prove* that $Br_2$ is the limiting reactant, the following calculation is needed:

$$0.03129 \text{ mol } Br_2 \times \frac{2 \text{ mol Na}}{1 \text{ mol } Br_2} = 0.06258 \text{ mol Na}$$

Clearly, there is more Na than this present, so $Br_2$ limits the reaction extent and the amount of NaBr formed.

$$0.03129 \text{ mol } Br_2 \times \frac{2 \text{ mol NaBr}}{1 \text{ mol } Br_2} = 0.06258 \text{ mol NaBr}$$

$$0.06258 \text{ mol NaBr} \times \frac{102.9 \text{ g NaBr}}{1 \text{ mol NaBr}} = 6.4 \text{ g NaBr}$$

b. $Zn(s) + CuSO_4(aq) \rightarrow ZnSO_4(aq) + Cu(s)$

molar masses: Zn, 65.38 g; Cu, 63.55 g;
$ZnSO_4$, 161.4 g; $CuSO_4$, 159.6 g

$$5.0 \text{ g Zn} \times \frac{1 \text{ mol Zn}}{65.38 \text{ g Zn}} = 0.07648 \text{ mol Zn}$$

$$5.0 \text{ g } CuSO_4 \times \frac{1 \text{ mol } CuSO_4}{159.6 \text{ g } CuSO_4} = 0.03132 \text{ mol } CuSO_4$$

Since the coefficients of Zn and $CuSO_4$ are the *same* in the balanced chemical equation, an equal number of moles of Zn and $CuSO_4$ would be needed for complete reaction. Since there is less $CuSO_4$ present, $CuSO_4$ must clearly be the limiting reactant.

$$0.03132 \text{ mol } CuSO_4 \times \frac{1 \text{ mol } ZnSO_4}{1 \text{ mol } CuSO_4} = 0.03132 \text{ mol } ZnSO_4$$

$$0.03132 \text{ mol } ZnSO_4 \times \frac{161.4 \text{ g } ZnSO_4}{1 \text{ mol } ZnSO_4} = 5.1 \text{ g } ZnSO_4$$

$$0.03132 \text{ mol } CuSO_4 \times \frac{1 \text{ mol Cu}}{1 \text{ mol } CuSO_4} = 0.03132 \text{ mol Cu}$$

$$0.03132 \text{ mol Cu} \times \frac{63.55 \text{ g Cu}}{1 \text{ mol Cu}} = 2.0 \text{ g Cu}$$

c. $NH_4Cl(aq) + NaOH(aq) \rightarrow NH_3(g) + H_2O(l) + NaCl(aq)$

molar masses: $NH_4Cl$, 53.49 g; NaOH, 40.00 g; $NH_3$, 17.03 g
$H_2O$, 18.02 g; NaCl, 58.44 g

$$5.0 \text{ g } NH_4Cl \times \frac{1 \text{ mol } NH_4Cl}{53.49 \text{ g } NH_4Cl} = 0.09348 \text{ mol } NH_4Cl$$

$$5.0 \text{ g NaOH} \times \frac{1 \text{ mol NaOH}}{40.00 \text{ g NaOH}} = 0.1250 \text{ mol NaOH}$$

Since the coefficients of $NH_4Cl$ and NaOH are both *one* in the balanced chemical equation for the reaction, an equal number of moles of $NH_4Cl$ and NaOH would be needed for complete reaction. Since there is less $NH_4Cl$ present, $NH_4Cl$ must be the limiting reactant.

Since the coefficients of the products in the balanced chemical equation are also all *one*, if 0.09348 mol of $NH_4Cl$ (the limiting reactant) reacts completely, then 0.09348 mol of each product will be formed.

$$0.09348 \text{ mol } NH_3 \times \frac{17.03 \text{ g } NH_3}{1 \text{ mol } NH_3} = 1.6 \text{ g } NH_3$$

$$0.09348 \text{ mol } H_2O \times \frac{18.02 \text{ g } H_2O}{1 \text{ mol } H_2O} = 1.7 \text{ g } H_2O$$

$$0.09348 \text{ mol NaCl} \times \frac{58.44 \text{ g NaCl}}{1 \text{ mol NaCl}} = 5.5 \text{ g NaCl}$$

d. $Fe_2O_3(s) + 3CO(g) \rightarrow 2Fe(s) + 3CO_2(g)$

molar masses: $Fe_2O_3$, 159.7 g; CO, 28.01 g
Fe, 55.85 g; $CO_2$, 44.01 g

$$5.0 \text{ g } Fe_2O_3 \times \frac{1 \text{ mol } Fe_2O_3}{159.7 \text{ g } Fe_2O_3} = 0.03131 \text{ mol } Fe_2O_3$$

$$5.0 \text{ g CO} \times \frac{1 \text{ mol CO}}{28.01 \text{ g CO}} = 0.1785 \text{ mol CO}$$

Because there is considerably less $Fe_2O_3$ than CO on a mole basis, let's see if $Fe_2O_3$ is the limiting reactant.

$$0.03131 \text{ mol } Fe_2O_3 \times \frac{3 \text{ mol CO}}{1 \text{ mol } Fe_2O_3} = 0.09393 \text{ mol CO}$$

Since there is 0.1785 mol of CO present, but we have determined that only 0.09393 mol CO would be needed to react with all the $Fe_2O_3$ present, then $Fe_2O_3$ must be the limiting reactant. CO is present in excess.

$$0.03131 \text{ mol } Fe_2O_3 \times \frac{2 \text{ mol Fe}}{1 \text{ mol } Fe_2O_3} = 0.06262 \text{ mol Fe}$$

$$0.06262 \text{ mol Fe} \times \frac{55.85 \text{ g Fe}}{1 \text{ mol Fe}} = 3.5 \text{ g Fe}$$

$$0.03131 \text{ mol } Fe_2O_3 \times \frac{3 \text{ mol } CO_2}{1 \text{ mol } Fe_2O_3} = 0.09393 \text{ mol } CO_2$$

$$0.09393 \text{ mol } CO_2 \times \frac{44.01 \text{ g } CO_2}{1 \text{ mol } CO_2} = 4.1 \text{ g } CO_2$$

48. a. $C_2H_5OH(l) + 3O_2(g) \rightarrow \mathbf{2CO_2(g)} + 3H_2O(l)$

molar masses: $C_2H_5OH$, 46.07 g; $O_2$, 32.00 g; $CO_2$, 44.01 g

$$25.0 \text{ g } C_2H_5OH \times \frac{1 \text{ mol } C_2H_5OH}{46.07 \text{ g } C_2H_5OH} = 0.5427 \text{ mol } C_2H_5OH$$

$$25.0 \text{ g } O_2 \times \frac{1 \text{ mol } O_2}{32.00 \text{ g } O_2} = 0.7813 \text{ mol } O_2$$

Since there is less $C_2H_5OH$ present on a mole basis, see if this substance is the limiting reactant.

$$0.5427 \text{ mol } C_2H_5OH \times \frac{3 \text{ mol } O_2}{1 \text{ mol } C_2H_5OH} = 1.6281 \text{ mol } O_2$$

From the above calculation, $C_2H_5OH$ must *not* be the limiting reactant (even though there is a smaller number of moles of $C_2H_5OH$ present) since more oxygen than is present would be required to react completely with the $C_2H_5OH$ present. Oxygen is the limiting reactant.

$$0.7813 \text{ mol } O_2 \times \frac{2 \text{ mol } CO_2}{3 \text{ mol } O_2} = 0.5209 \text{ mol } CO_2$$

$$0.5209 \text{ mol } CO_2 \times \frac{44.01 \text{ g } CO_2}{1 \text{ mol } CO_2} = 22.9 \text{ g } CO_2$$

b. $N_2(g) + O_2(g) \rightarrow \mathbf{2NO}(g)$

molar masses: $N_2$, 28.02 g; $O_2$, 32.00 g; NO, 30.01 g

$$25.0 \text{ g } N_2 \times \frac{1 \text{ mol } N_2}{28.02 \text{ g } N_2} = 0.8922 \text{ mol } N_2$$

$$25.0 \text{ g } O_2 \times \frac{1 \text{ mol } O_2}{32.00 \text{ g } O_2} = 0.7813 \text{ mol } O_2$$

Since the coefficients of $N_2$ and $O_2$ are the *same* in the balanced chemical equation for the reaction, an equal number of moles of each substance would be necessary for complete reaction. Since there is less $O_2$ present on a mole basis, $O_2$ must be the limiting reactant.

$$0.7813 \text{ mol } O_2 \times \frac{2 \text{ mol NO}}{1 \text{ mol } O_2} = 1.5626 \text{ mol NO}$$

$$1.5626 \text{ mol NO} \times \frac{30.01 \text{ g NO}}{1 \text{ mol NO}} = 46.9 \text{ g NO}$$

c. $2NaClO_2(aq) + Cl_2(g) \rightarrow 2ClO_2(g) + \mathbf{2NaCl}(aq)$

molar masses: $NaClO_2$, 90.44 g; $Cl_2$, 70.90 g; NaCl, 58.44 g

$$25.0 \text{ g } NaClO_2 \times \frac{1 \text{ mol } NaClO_2}{90.44 \text{ g } NaClO_2} = 0.2764 \text{ mol } NaClO_2$$

$$25.0 \text{ g } Cl_2 \times \frac{1 \text{ mol } Cl_2}{70.90 \text{ g } Cl_2} = 0.3526 \text{ mol } Cl_2$$

See if $NaClO_2$ is the limiting reactant.

$$0.2764 \text{ mol } NaClO_2 \times \frac{1 \text{ mol } Cl_2}{2 \text{ mol } NaClO_2} = 0.1382 \text{ mol } Cl_2$$

Since 0.2764 mol of $NaClO_2$ would require only 0.1382 mol $Cl_2$ to react completely (and since we have more than this amount of $Cl_2$), then $NaClO_2$ must indeed be the limiting reactant.

$$0.2764 \text{ mol } NaClO_2 \times \frac{2 \text{ mol } NaCl}{2 \text{ mol } NaClO_2} = 0.2764 \text{ mol } NaCl$$

$$0.2764 \text{ mol } NaCl \times \frac{58.44 \text{ g } NaCl}{1 \text{ mol } NaCl} = 16.2 \text{ g } NaCl$$

d. $3H_2(g) + N_2(g) \rightarrow 2NH_3(g)$

molar masses: $H_2$, 2.016 g; $N_2$, 28.02 g; $NH_3$, 17.03 g

$$25.0 \text{ g } H_2 \times \frac{1 \text{ mol } H_2}{2.016 \text{ g } H_2} = 12.40 \text{ mol } H_2$$

$$25.0 \text{ g } N_2 \times \frac{1 \text{ mol } N_2}{28.02 \text{ g } N_2} = 0.8922 \text{ mol } N_2$$

See if $N_2$ is the limiting reactant.

$$0.8922 \text{ mol } N_2 \times \frac{3 \text{ mol } H_2}{1 \text{ mol } N_2} = 2.677 \text{ mol } H_2$$

$N_2$ is clearly the limiting reactant, since there is 12.40 mol $H_2$ present (a large excess).

$$0.8922 \text{ mol } N_2 \times \frac{2 \text{ mol } NH_3}{1 \text{ mol } N_2} = 1.7844 \text{ mol } NH_3$$

$$1.7844 \text{ mol } NH_3 \times \frac{17.03 \text{ g } NH_3}{1 \text{ mol } NH_3} = 30.4 \text{ g } NH_3$$

50. a. $CO(g) + 2H_2(g) \rightarrow CH_3OH(l)$

CO is the limiting reactant; 11.4 mg $CH_3OH$

b. $2Al(s) + 3I_2(s) \rightarrow 2AlI_3(s)$

$I_2$ is the limiting reactant; 10.7 mg $AlI_3$

c. $Ca(OH)_2(aq) + 2HBr(aq) \rightarrow CaBr_2(aq) + 2H_2O(l)$

HBr is the limiting reactant; 12.3 mg $CaBr_2$; 2.22 mg $H_2O$

d. $2Cr(s) + 2H_3PO_4(aq) \rightarrow 2CrPO_4(s) + 3H_2(g)$

$H_3PO_4$ is the limiting reactant; 15.0 mg $CrPO_4$; 0.309 mg $H_2$

52. $2NH_3(g) + CO_2(g) \rightarrow CN_2H_4O(s) + H_2O(l)$

molar masses: $NH_3$, 17.03 g; $CO_2$, 44.01 g; $CN_2H_4O$, 60.06 g

$$100.\text{ g } NH_3 \times \frac{1 \text{ mol } NH_3}{17.03 \text{ g } NH_3} = 5.872 \text{ mol } NH_3$$

$$100.\text{ g } CO_2 \times \frac{1 \text{ mol } CO_2}{44.02 \text{ g } CO_2} = 2.272 \text{ mol } CO_2$$

See if $CO_2$ is the limiting reactant.

$$2.272 \text{ mol } CO_2 \times \frac{2 \text{ mol } NH_3}{1 \text{ mol } CO_2} = 4.544 \text{ mol } NH_3$$

$CO_2$ is indeed the limiting reactant.

$$2.272 \text{ mol } CO_2 \times \frac{1 \text{ mol } CN_2H_4O}{1 \text{ mol } CO_2} = 2.272 \text{ mol } CN_2H_4O$$

$$2.272 \text{ mol } CN_2H_4O \times \frac{60.03 \text{ g } CN_2H_4O}{1 \text{ mol } CN_2H_4O} = 136 \text{ g } CN_2H_4O$$

54. $N_2H_4(l) + O_2(g) \rightarrow N_2(g) + 2H_2O(g)$

molar masses: $N_2H_4$, 32.05 g; $O_2$, 32.00 g; $N_2$, 28.02 g; $H_2O$, 18.02 g

$$20.0 \text{ g } N_2H_4 \times \frac{1 \text{ mol } N_2H_4}{32.05 \text{ g } N_2H_4} = 0.624 \text{ mol } N_2H_4$$

$$20.0 \text{ g } O_2 \times \frac{1 \text{ mol } O_2}{32.00 \text{ g } O_2} = 0.625 \text{ mol } O_2$$

The two reactants are present in very nearly the required ratio for complete reaction (due to the 1:1 stoichiometry of the reaction and the very similar molar masses of the substances). We will consider $N_2H_4$ as the limiting reactant in the following calculations.

$$0.624 \text{ mol } N_2H_4 \times \frac{1 \text{ mol } N_2}{1 \text{ mol } N_2H_4} = 0.624 \text{ mol } N_2$$

$$0.624 \text{ mol } N_2 \times \frac{28.02 \text{ g } N_2}{1 \text{ mol } N_2} = 17.5 \text{ g } N_2$$

$$0.624 \text{ mol } N_2H_4 \times \frac{2 \text{ mol } H_2O}{1 \text{ mol } N_2H_4} = 1.248 \text{ mol } H_2O = 1.25 \text{ mol } H_2O$$

$$1.248 \text{ mol } H_2O \times \frac{18.02 \text{ g } H_2O}{1 \text{ mol } H_2O} = 22.5 \text{ g } H_2O$$

56. Total quantity of $H_2S$ = $50. \text{ L} \times \frac{1.5 \times 10^{-5} \text{ g}}{1 \text{ L}} = 7.5 \times 10^{-4} \text{ g } H_2S$

$8H_2S(aq) + 8Cl_2(aq) \rightarrow 16HCl(aq) + S_8(s)$

molar masses: $H_2S$, 34.08 g; $Cl_2$, 70.90 g; $S_8$, 256.5 g

$$7.5 \times 10^{-4} \text{ g } H_2S \times \frac{1 \text{ mol } H_2S}{34.08 \text{ g } H_2S} = 2.20 \times 10^{-5} \text{ mol } H_2S$$

$$1.0 \text{ g } Cl_2 \times \frac{1 \text{ mol } Cl_2}{70.90 \text{ g } Cl_2} = 1.41 \times 10^{-2} \text{ mol } Cl_2$$

There is a large excess of chlorine present, compared to the amount of $Cl_2$ that would be needed to react with all the $H_2S$ present in the water sample: $H_2S$ is the limiting reactant for the process.

$$2.20 \times 10^{-5} \text{ mol } H_2S \times \frac{1 \text{ mol } S_8}{8 \text{ mol } H_2S} = 2.75 \times 10^{-6} \text{ mol } S_8$$

$$2.75 \times 10^{-6} \text{ mol } S_8 \times \frac{256.5 \text{ g } S_8}{1 \text{ mol } S_8} = 7.1 \times 10^{-4} \text{ g } S_8 \text{ removed}$$

58. $SiO_2(s) + C(s) \rightarrow 2CO(g) + SiC(s)$

molar masses: $SiO_2$, 60.09 g; SiC, 40.1

$1.0 \text{ kg} = 1.0 \times 10^3 \text{ g}$

$$1.0 \times 10^3 \text{ g } SiO_2 \times \frac{1 \text{ mol } SiO_2}{60.09 \text{ g } SiO_2} = 16.64 \text{ mol } SiO_2$$

From the balanced chemical equation, if 16.64 mol of $SiO_2$ were to react completely (an excess of carbon is present), then 16.64 mol of SiC should be produced (the coefficients of $SiO_2$ and SiC are the same).

$$16.64 \text{ mol SiC} \times \frac{40.01 \text{ g SiC}}{1 \text{ mol SiC}} = 6.7 \times 10^2 \text{ g SiC} = 0.67 \text{ kg SiC}$$

## 10.5 Percent Yield

60. If the reaction is performed in a solvent, the product may have a substantial solubility in the solvent; the reaction may come to equilibrium before the full yield of product is achieved (see Chapter 16); loss of product may occur through operator error.

62. $12.5 \text{ g theory} \times \dfrac{40 \text{ g actual}}{100 \text{ g theory}} = 5 \text{ g}$

64. $SiO_2(s) + 2C(s) \rightarrow Si(s) + 2CO(g)$

molar masses: $SiO_2$, 60.09g; C, 12.01 g; Si, 28.09 g

$100. \text{ kg} = 1.00 \times 10^5 \text{ g}$

$$1.00 \times 10^5 \text{ g } SiO_2 \times \frac{1 \text{ mol } SiO_2}{60.09 \text{ g } SiO_2} = 1.664 \times 10^3 \text{ mol } SiO_2$$

$$1.00 \times 10^5 \text{ g C} \times \frac{1 \text{ mol C}}{12.00 \text{ g C}} = 8.326 \times 10^3 \text{ mol C}$$

$SiO_2$ is the limiting reactant.

$$1.664 \times 10^3 \text{ mol } SiO_2 \times \frac{1 \text{ mol Si}}{1 \text{ mol } SiO_2} = 1.664 \times 10^3 \text{ mol Si}$$

$$1.664 \times 10^3 \text{ mol Si} \times \frac{28.09 \text{ g Si}}{1 \text{ mol Si}} = 4.674 \times 10^4 \text{ g Si} = 46.7 \text{ kg Si}$$

$$\text{Percent yield} = \frac{\text{actual yield}}{\text{theoretical yield}} \times 100 = \frac{17.2 \text{ kg}}{46.7 \text{ kg}} \times 100 = 36.8\%$$

66. $CaCO_3(s) + 2HCl(g) \rightarrow CaCl_2(s) + CO_2(g) + H_2O(g)$

molar masses: $CaCO_3$, 100.1 g; HCl, 36.46 g; $CaCl_2$, 111.0 g

$$155 \text{ g } CaCO_3 \times \frac{1 \text{ mol } CaCO_3}{100.1 \text{ g } CaCO_3} = 1.548 \text{ mol } CaCO_3$$

$$250. \text{ g HCl} \times \frac{1 \text{ mol HCl}}{36.46 \text{ g HCl}} = 6.857 \text{ mol HCl}$$

$CaCO_3$ is the limiting reactant.

$$1.548 \text{ mol } CaCO_3 \times \frac{1 \text{ mol } CaCl_2}{1 \text{ mol } CaCO_3} = 1.548 \text{ mol } CaCl_2$$

$$1.548 \text{ mol } CaCl_2 \times \frac{111.0 \text{ g } CaCl_2}{1 \text{ mol } CaCl_2} = 172 \text{ g } CaCl_2$$

$$\text{Percent yield} = \frac{\text{actual yield}}{\text{theoretical yield}} \times 100 = \frac{142\ \text{g}}{172\ \text{g}} \times 100 = 82.6\%$$

## Additional Problems

68. $NaCl(aq) + NH_3(aq) + H_2O(l) + CO_2(s) \rightarrow NH_4Cl(aq) + NaHCO_3(s)$

molar masses: $NH_3$, 17.03 g; $CO_2$, 44.01 g; $NaHCO_3$, 84.01 g

$$10.0\ \text{g}\ NH_3 \times \frac{1\ \text{mol}\ NH_3}{17.03\ \text{g}\ NH_3} = 0.5872\ \text{mol}\ NH_3$$

$$15.0\ \text{g}\ CO_2 \times \frac{1\ \text{mol}\ CO_2}{44.01\ \text{g}\ CO_2} = 0.3408\ \text{mol}\ CO_2$$

$CO_2$ is the limiting reactant.

$$0.3408\ \text{mol}\ CO_2 \times \frac{1\ \text{mol}\ NaHCO_3}{1\ \text{mol}\ CO_2} = 0.3408\ \text{mol}\ NaHCO_3$$

$$0.3408\ \text{mol}\ NaHCO_3 \times \frac{84.01\ \text{g}\ NaHCO_3}{1\ \text{mol}\ NaHCO_3} = 28.6\ \text{g}\ NaHCO_3$$

70. $C_6H_{12}O_6(s) + 6O_2(g) \rightarrow 6CO_2(g) + 6H_2O(g)$

molar masses: glucose, 180.2 g; $CO_2$, 44.01 g

$$1.00\ \text{g glucose} \times \frac{1\ \text{mol glucose}}{180.2\ \text{g glucose}} = 5.549 \times 10^{-3}\ \text{mol glucose}$$

$$5.549 \times 10^{-3}\ \text{mol glucose} \times \frac{6\ \text{mol}\ CO_2}{1\ \text{mol glucose}} = 3.32 \times 10^{-2}\ \text{mol}\ CO_2$$

$$3.32 \times 10^{-2}\ \text{mol}\ CO_2 \times \frac{44.01\ \text{g}\ CO_2}{1\ \text{mol}\ CO_2} = 1.47\ \text{g}\ CO_2$$

72. $Pb^{2+}(aq) + CrO_4^{2-}(aq) \rightarrow PbCrO_4(s)$

millimolar ionic masses: $Pb^{2+}$, 207.2 mg; $K_2CrO_4$, 194.2 mg

$$15 \text{ mg } Pb^{2+} \times \frac{1 \text{ mmol } Pb^{2+}}{207.2 \text{ mg } Pb^{2+}} = 0.07239 \text{ mmol } Pb^{2+}$$

$$0.07239 \text{ mmol } Pb^{2+} \times \frac{1 \text{ mmol } CrO_4^{2-}}{1 \text{ mmol } Pb^{2+}} = 0.07239 \text{ mmol } CrO_4^{2-}$$

$$0.07239 \text{ mmol } CrO_4^{2-} = 0.07239 \text{ mmol } K_2CrO_4$$

$$0.07239 \text{ mmol } CrO_4^{2-} \times \frac{194.2 \text{ mg } K_2CrO_4}{1 \text{ mmol } K_2CrO_4} = 14 \text{ mg } K_2CrO_4$$

# Chapter Eleven Modern Atomic Theory

## 11.1 Electromagnetic Radiation and Energy

2. The wavelength ($\lambda$) represents the distance between two corresponding points (peaks, troughs, etc.) on successive cycles of a wave.

4. The molecule moves or rotates in space at a higher speed, and atoms in the molecule vibrate more vigorously.

6. $3.00 \times 10^8$ m/sec

## 11.2 The Energy Levels of Hydrogen

8. photon

10. lower (the frequency of red light is lower than that of blue light)

12. Only *certain* energy levels are allowed to the electron in the hydrogen atom. These levels correspond to definite, distinct *energies*. When an electron moves from one allowed level to another, a characteristic photon of radiation is emitted.

14. lower energy state [often the ground (lowest energy) state]

16. Light is emitted from the hydrogen atom only at certain fixed wavelengths. If the energy levels of hydrogen were *continuous*, a hydrogen atom would emit energy at all possible wavelengths.

## 11.3 The Bohr Model of the Atom

18. According to Bohr, electrons move in discrete, fixed circular *orbits* around the nucleus. If the wavelength of the applied energy corresponds to the *difference in energy* between the two orbits, the atom absorbs a photon and the electron moves to a larger orbit.

20. Bohr's theory *explained* the experimentally *observed* line spectrum of hydrogen *exactly*. Bohr's theory was ultimately discarded because when attempts were made to extend the theory to atoms other than hydrogen, the calculated properties did *not* correspond closely to experimental measurements.

## 11.4 The Wave Mechanical Model of the Atom

22. An orbit represents a definite, exact circular pathway around the nucleus in which an electron can be found. An orbital represents a region of space in which there is a high probability of finding the electron.

24. Any experiment which sought to measure the exact location of an electron (such as shooting a beam of light at it) would cause the electron to move. Any measurement made would necessitate the application or removal of energy, which would disturb the electron from where it had been before the measurement.

## 11.5 The Hydrogen Orbitals

26. The principal energy levels represent sets of orbitals at a particular average distance from the nucleus and a particular average energy in which electrons may reside. With the Bohr theory, the nucleus was surrounded by a series of circular orbits of fixed radius in which electrons moved.

28. The 2*p* orbitals have two lobes and are sometimes described as having a "dumbbell" shape. The individual 2*p* orbitals ($2p_x$, $2p_y$, $2p_z$) are alike in shape and energy; they differ only in the direction in which the lobes of the orbital are oriented.

30. decreases (as an electron moves to a higher-number level, the electron's mean distance from the nucleus increases, thereby decreasing the attractive force between the electron and the nucleus).

32. The third principal energy level of hydrogen is divided into *three* sublevels (3*s*, 3*p*, and 3*d*); there is a *single* 3*s* orbital; there is a set of *three* 3*p* orbitals; there is a set of *five* 3*d* orbitals.

## 11.6 The Wave Mechanical Model: A Summary

34. The two electrons must have *opposite* (paired) spins.

36. increases

38. probability

40. a. incorrect; there is only an *s* subshell in the first principal energy level

    b. correct

    c. incorrect; the third principal energy level has only *s*, *p*, and *d* subshells possible

    d. correct

    e. correct

    f. correct

## 11.7 Electron Arrangements in the First Eighteen Atoms

42. The three 2*p* orbitals of carbon are of the *same energy*; by occupying different orbitals of the same energy, repulsion between electrons is minimized.

44. The elements in a given vertical column of the periodic table have the same valence electron configuration.

46. a. $1s^2\ 2s^2\ 2p^6\ 3s^2\ 3p^5$
    b. $1s^2\ 2s^2\ 2p^4$
    c. $1s^2\ 2s^2\ 2p^6\ 3s^2\ 3p^1$
    d. $1s^2\ 2s^1$

48. a. $1s^2\ 2s^2\ 2p^2$
    b. $1s^2\ 2s^2\ 2p^6\ 3s^2\ 3p^3$
    c. $1s^2\ 2s^2\ 2p^6\ 3s^2\ 3p^4$
    d. $1s^2\ 2s^2\ 2p^1$

50. a. (↑↓) (↑↓) (↑↓)(↑↓)(↑↓) (↑↓) (↑↓)(↑↓)(↑↓) (↑↓) (↑ )( )( )( )( )
    1*s* 2*s* 2*p* 3*s* 3*p* 4*s* 3*d*

    b. (↑↓) (↑↓) (↑↓)(↑↓)(↑↓) (↑↓) (↑↓)(↑ )(↑ )
    1*s* 2*s* 2*p* 3*s* 3*p*

    c. (↑↓) (↑↓) (↑↓)(↑↓)(↑↓) (↑↓) (↑↓)(↑↓)(↑↓) (↑ )
    1*s* 2*s* 2*p* 3*s* 3*p* 4*s*

    d. (↑↓) (↑↓) (↑ )(↑ )(↑ )
    1*s* 2*s* 2*p*

52. a. five (2*s*, 2*p*)
    b. seven (2*s*, 2*p*)
    c. one (3*s*)
    d. three (3*s*, 3*p*)

## 11.8 Electron Configurations and the Periodic Table

54. transition metals

56. a. [Kr] $5s^2\ 4d^2$
    b. [Kr] $5s^2\ 4d^{10}\ 5p^5$
    c. [Ar] $4s^2\ 3d^{10}\ 4p^2$
    d. [Xe] $6s^1$

58. a. [Ar] $4s^2\ 3d^2$

    b. [Ar] $4s^2\ 3d^{10}\ 4p^4$

    c. [Kr] $5s^2\ 4d^{10}\ 5p^3$

    d. [Kr] $5s^2$

60. The number of $4d$ electrons can be determined from the *position* of the element in the $4d$ row of the periodic table.

    a. one

    b. two

    c. zero

    d. ten

62. The *position* of the element (both in terms of the vertical column and the horizontal row) tells you which set of orbitals is being filled last. See Figure 11.27 for details.

    a. $3d$

    b. $4d$

    c. $5f$

    d. $4p$

64. a. [Ar] $4s^2\ 3d^8$

    b. [Kr] $5s^2\ 4d^3$ (actually [Kr] $5s^1\ 4d^4$ for reasons beyond text)

    c. [Xe] $6s^2\ 4f^{14}\ 5d^2$

    d. [Xe] $6s^2\ 4f^{14}\ 5d^{10}\ 6p^5$

## 11.9 Atomic Properties and the Periodic Table

66. The metallic elements *lose* electrons and form *positive* ions (cations); the nonmetallic elements *gain* electrons and form *negative* ions (anions).

68. All exist as *diatomic* molecules ($F_2$, $Cl_2$, $Br_2$, $I_2$); all are *non*metals; all have relatively high electronegativities; all form 1– ions in reacting with metallic elements.

70. Elements at the *left* of a period (horizontal row) lose electrons more readily; at the left of a period (given principal energy level) the nuclear charge is the smallest and the electrons are least tightly held.

72. metals-low; nonmetals-high

74. The *nuclear charge* increases from left to right within a period, pulling progressively more tightly on the valence electrons.

76. a. B and Al are both very reactive

    b. Na

    c. F

78. a. Ba

    b. Mg

    c. Be

## Additional Problems

80. speed of light

82. photons

84. quantized

86. orbital

88. transition metal

90. spins

92. a. $1s^2\ 2s^2\ 2p^6\ 3s^2\ 3p^6\ 4s^1$ [Ar] $4s^1$

| (↑↓) | (↑↓) | (↑↓)(↑↓)(↑↓) | (↑↓) | (↑↓)(↑↓)(↑↓) | (↑ ) |
|---|---|---|---|---|---|
| 1*s* | 2*s* | 2*p* | 3*s* | 3*p* | 4*s* |

b. $1s^2\ 2s^2\ 2p^6\ 3s^2\ 3p^6\ 4s^2\ 3d^2$ [Ar] $4s^2\ 3d^2$

| (↑↓) | (↑↓) | (↑↓)(↑↓)(↑↓) | (↑↓) | (↑↓)(↑↓)(↑↓) | (↑↓) | (↑ )(↑ )( )( )( ) |
|---|---|---|---|---|---|---|
| 1*s* | 2*s* | 2*p* | 3*s* | 3*p* | 4*s* | 3*d* |

c. $1s^2\ 2s^2\ 2p^6\ 3s^2\ 3p^2$ [Ne] $3s^2\ 3p^2$

| (↑↓) | (↑↓) | (↑↓)(↑↓)(↑↓) | (↑↓) | (↑ )(↑ )( ) |
|---|---|---|---|---|
| 1*s* | 2*s* | 2*p* | 3*s* | 3*p* |

d. $1s^2\ 2s^2\ 2p^6\ 3s^2\ 3p^6\ 4s^2\ 3d^6$ [Ar] $4s^2\ 3d^6$

| (↑↓) | (↑↓) | (↑↓)(↑↓)(↑↓) | (↑↓) | (↑↓)(↑↓)(↑↓) | (↑↓) | (↑↓)(↑ )(↑ )(↑ )(↑ ) |
|---|---|---|---|---|---|---|
| 1*s* | 2*s* | 2*p* | 3*s* | 3*p* | 4*s* | 3*d* |

e. $1s^2\ 2s^2\ 2p^6\ 3s^2\ 3p^6\ 4s^2\ 3d^{10}$ [Ar] $4s^2\ 3d^{10}$

(↑↓) (↑↓) (↑↓)(↑↓)(↑↓) (↑↓) (↑↓)(↑↓)(↑↓) (↑↓) (↑↓)(↑↓)(↑↓)(↑↓)(↑↓)
$1s$ $2s$ $2p$ $3s$ $3p$ $4s$ $3d$

94. a. $ns^2$

b. $ns^2\ np^5$

c. $ns^2\ np^4$

d. $ns^1$

e. $ns^2\ np^4$

96. a. $$\lambda = \frac{h}{mv}$$

$$\lambda = \frac{6.63 \times 10^{-34}\ \text{J s}}{(9.11 \times 10^{-31}\ \text{kg})[0.90(3.00 \times 10^{8}\ \text{m s}^{-1})]}$$

$\lambda = 2.7 \times 10^{-12}$ m (0.0027 nm)

b. $4.4 \times 10^{-34}$ m

c. $2 \times 10^{-35}$ m

The wavelengths for the ball and the person are *infinitesimally small*, whereas the wavelength for the electron is nearly the same order of magnitude as the diameter of a typical atom.

# Chapter Twelve Chemical Bonding

## 12.1 Types of Chemical Bonds

2. break

4. covalent

6 polar

## 12.2 Electronegativity

8 electronegativity

10. difference

12. In each case, the element *higher up* within a group of the periodic table has the higher electronegativity.

    a. Be

    b. N

    c. F

14. Generally, covalent bonds between atoms of *different* elements are *polar*.

    a. polar covalent

    b. covalent

    c. covalent

    d. polar covalent

16. For a bond to be polar covalent, the atoms involved in the bond must have different electronegativities (must be of different elements).

    a. polar covalent (different elements)

    b. *non*polar covalent (two atoms of the same element)

    c. polar covalent (different elements)

    d. *non*polar covalent (atoms of the same element)

18. The *degree* of polarity of a polar covalent bond is indicated by the magnitude of the difference in electronegativities of the elements involved: the larger the difference in electronegativity, the more polar is the bond. Electronegativity differences are given in parentheses below:

a. H–O (1.4); H–N (0.9); the H–O bond is more polar.

b. H–N (0.9); H–F (1.9); the H–F bond is more polar.

c. H–O (1.4); H–F (1.9); the H–F bond is more polar.

d. H–O (1.4); H–Cl (0.9); the H–O bond is more polar.

20. Electronegativity differences are given in parentheses.

a. N–P (0.9); N–O (0.5); the N–P bond is more polar.

b. N–C (0.5); N–O (0.5); the bonds are of the same polarity.

c. N–S (0.5); N–C (0.5); the bonds are of the same polarity.

d. N–F (1.0); N–S (0.5); the N–F bond is more polar.

## 12.3 Bond Polarity and Dipole Moments

22. The presence of strong bond dipoles and a large overall dipole moment in water make it a very polar substance overall. Among those properties of water that are dependent on its dipole moment are its freezing point, melting point, vapor pressure, and its ability to dissolve many substances.

24. In a diatomic molecule containing two different elements, the more electronegative atom will be the negative end of the molecule, and the *less* electronegative atom will be the positive end.

a. hydrogen

b. iodine

c. nitrogen

26. In the figures, the arrow points toward the more electronegative atom.

a. N–Cl The atoms have very nearly the same electronegativity, so there is a very small, if any, dipole moment.

b. $\delta+$ P→N $\delta-$

c. $\delta+$ S→N $\delta-$

d. $\delta+$ C→N $\delta-$

28. In the figures, the arrow points toward the more electronegative atom.

    a. $\delta+$ P→S $\delta-$

    b. $\delta+$ S→O $\delta-$

    c. $\delta+$ S→N $\delta-$

    d. $\delta+$ S→Cl $\delta-$

## 12.4 Stable Electron Configurations and Charges on Ions

30. previous

32. Atoms in covalent molecules gain a configuration like that of a noble gas by sharing one or more pairs of electrons between atoms: such shared pairs of electrons "belong" to each of the atoms of the bond at the same time. In ionic bonding, one atom completely gives over one or more electrons to another atom, and then the resulting ions behave independently of one another (they are not "attached" to each other as in the case of a covalent bond, although they are attracted to each other).

34. a. Al $1s^2\ 2s^2\ 2p^6\ 3s^2\ 3p^1$

    $Al^{3+}$ $1s^2\ 2s^2\ 2p^6$

    Ne has the same configuration as $Al^{3+}$.

    b. Br $1s^2\ 2s^2\ 2p^6\ 3s^2\ 3p^6\ 4s^2\ 3d^{10}\ 4p^5$

    $Br^-$ $1s^2\ 2s^2\ 2p^6\ 3s^2\ 3p^6\ 4s^2\ 3d^{10}\ 4p^6$

    Kr has the same configuration as $Br^-$.

    c. Ca $1s^2\ 2s^2\ 2p^6\ 3s^2\ 3p^6\ 4s^2$

    $Ca^{2+}$ $1s^2\ 2s^2\ 2p^6\ 3s^2\ 3p^6$

    Ar has the same configuration as $Ca^{2+}$.

    d. Li $1s^2\ 2s^1$

    $Li^+$ $1s^2$

    He has the same configuration as $Li^+$.

    e. F $1s^2\ 2s^2\ 2p^5$

    $F^-$ $1s^2\ 2s^2\ 2p^6$

    Ne has the same configuration as $F^-$.

36. a. $Ba^{2+}$ (Ba has two electrons more than the noble gas Xe)
    b. $Rb^{+}$ (Rb has one electon more than the noble gas Kr)
    c. $Al^{3+}$ (Al has three electrons more than the noble gas Ne)
    d. $O^{2-}$ (O has two electrons fewer than the noble gas Ne)

38. a. $Na_2Se$ Na has one electron more than a noble gas; Se has two electrons fewer than a noble gas.

    b. RbF Rb has one electron more than a noble gas; F has one electron less than a noble gas.

    c. $K_2Te$ K has one electron more than a noble gas; Te has two electrons fewer than a noble gas.

    d. BaSe Ba has two electrons more than a noble gas; Se has two electrons fewer than a noble gas.

    e. KAt K has one electron more than a noble gas; At has one electron less than a noble gas.

    f. FrCl Fr has one electron more than a noble gas; Cl has one electron less than a noble gas.

40. a. $Ca^{2+}$ [Ar]; $Br^{-}$ [Kr]
    b. $Al^{3+}$ [Ne]; $Se^{2-}$[Kr]
    c. $Sr^{2+}$ [Kr]; O [Ne]
    d. $K^{+}$ [Ar]; $S^{2-}$ [Ar]

## 12.5 Ionic Bonding and Structures of Ionic Compounds

42. An ionic solid such as NaCl basically consists of an array of alternating positively and negatively charged ions: that is, each positive ion has as its nearest neighbors a group of negative ions, and each negative ion has a group of positive ions surrounding it. In most ionic solids, the ions are packed as tightly as possible.

44. A polyatomic ion such as $SO_4^{2-}$ can be thought of as a covalently bonded *molecule* which carries a *net charge* (a polyatomic ion is an ion and can attract oppositely-charged ions).

46. Relative ionic sizes are indicated in Figure 12.9. Within a given horizontal row of the periodic chart, negative ions tend to be larger than positive ions because the negative ions contain a larger number of electrons in the valence shell.

    a. $F^-$

    b. $Cl^-$

    c. $O^{2-}$

    d. $I^-$

48. Relative ionic sizes are indicated in Figure 12.9.

    a. $Na^+$

    b. $Al^{3+}$

    c. $F^-$

    d. $Na^+$

## 12.6 Lewis Structures

50. When atoms form covalent bonds, they try to attain a valence electronic configuration similar to that of the nearest noble gas element. When the elements in the first few horizontal rows of the periodic table form covalent bonds, they will attempt to gain configurations similar to the noble gases helium (2 valence electrons, duet rule), and neon and argon (8 valence electrons, octet rule).

52. These elements attain a total of eight valence electrons, making the valence electron configurations similar to those of the noble gases Ne and Ar.

54. two

56. a. He:

    b. :B̤̈r·

    c. Sr:

    d. :N̤̈e:

    e. :Ï̤·

    f. Ra:

58. a. Si provides 4; each H provides 1; total valence electrons = 8
    b. S provides 6; each F provides 7; total valence electrons = 48
    c. each N provides 5; each O provides 6; total valence electrons = 34
    d. As provides 5; each I provides 7; total valence electrons = 26

60. a. $GeH_4$ Ge provides 4 valence electrons.
    Each H provides 1 valence electron.
    Total valence electrons = 8

```
     H
     |
  H—Ge—H
     |
     H
```

   b. ICl I provides 7 valence electrons.
    Cl provides 7 valence electrons.
    Total valence electrons = 14

```
  ..  ..
 :I—Cl:
  ..  ..
```

   c. $NI_3$ N provides 5 valence electrons.
    Each I provides 7 valence electrons.
    Total valence electrons = 26

```
  ..  ..  ..
 :I—N—I:
  ..  |   ..
     :I:
      ..
```

   d. $PF_3$ P provides 5 valence electrons.
    Each F provides 7 valence electrons.
    Total valence electrons = 26

```
  ..  ..  ..
 :F—P—F:
  ..  |   ..
     :F:
      ..
```

62. a. $N_2H_4$ Each N provides 5 valence electrons.
    Each H provides 1 valence electron.
    Total valence electrons = 14

```
    ..  ..
  H—N—N—H
    |   |
    H   H
```

b. $C_2H_6$

Each C provides 4 valence electrons.
Each H provides 1 valence electron.
Total valence electrons = 14

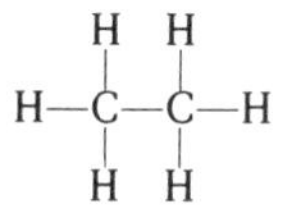

```
   H  H
   |  |
H—C—C—H
   |  |
   H  H
```

c. $NCl_3$

N provides 5 valence electrons.
Each Cl provides 7 valence electrons.
Total valence electrons = 26

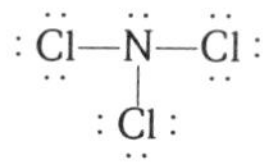

```
:Cl—N—Cl:
     |
    :Cl:
```

d. $SiCl_4$

Si provides 4 valence electrons.
Each Cl provides 7 valence electrons.
Total valence electrons = 32

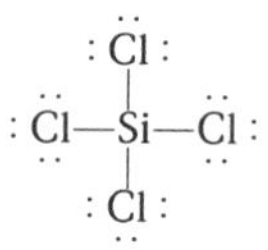

```
      :Cl:
       |
:Cl—Si—Cl:
       |
      :Cl:
```

64. a. $SO_2$

S provides 6 valence electrons.
Each O provides 6 valence electrons.
Total valence electrons = 18

```
O=S—O: ↔ :O—S=O
```

b. $N_2O$

Each N provides 5 valence electrons.
O provides 6 valence electrons.
Total valence electrons = 16

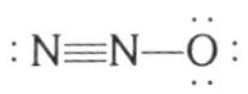

```
:N≡N—O:
```

c. $O_3$

Each O provides 6 valence electrons.
Total valence electrons = 18

```
O=O—O: ↔ :O—O=O
```

66. a. $NO_3^-$

N provides 5 valence electrons.
Each O provides 6 valence electrons.
The 1– charge means one additional valence electron.
Total valence electrons = 24

$$\left[O{=}N(-O){-}O\right]^- \leftrightarrow \left[O{-}N(-O){=}O\right]^- \leftrightarrow \left[O{-}N({=}O){-}O\right]^-$$

b. $CO_3^{2-}$

C provides 4 valence electrons.
Each O provides 6 valence electrons.
The 2– charge means two additional valence electrons.
Total valence electrons = 24

$$\left[O{=}C(-O){-}O\right]^{2-} \leftrightarrow \left[O{-}C(-O){=}O\right]^{2-} \leftrightarrow \left[O{-}C({=}O){-}O\right]^{2-}$$

c. $NH_4^+$

N provides 5 valence electrons.
Each H provides 1 valence electron.
The 1+ charge means one *less* valence electron.
Total valence electrons = 8

$$\left[H{-}N(-H)(-H){-}H\right]^+$$

68. a. $CN^-$

C provides 4 valence electrons.
N provides 5 valence electrons.
The 1– charge means one additional valence electron.
Total valence electrons = 10

$$[{:}C{\equiv}N{:}]^-$$

b. $HSO_4^-$

H provides 1 valence electron.
S provides 6 valence electrons.
Each O provides 6 valence electrons.
The 1– charge means one additional valence electron.
Total valence electrons = 32

$$\left[O{-}S(-O)(-O){-}O{:}H\right]^-$$

c. $N_3^-$ Each N provides 5 valence electrons.
The 1– charge means one additional valence electron.
Total valence electrons = 16

$$[:\ddot{N}=N=\ddot{N}:]^- \leftrightarrow [:N\equiv N-\ddot{\underset{..}{N}}:]^- \leftrightarrow [:\ddot{\underset{..}{N}}-N\equiv N:]^-$$

## 12.8 Molecular Structure

70. The geometric structure of $NH_3$ is that of a trigonal pyramid. The nitrogen atom of $NH_3$ is surrounded by four electron pairs (three are bonding, one is a lone pair). The H–N–H bond angle is somewhat less than 109.5° (due to the presence of the lone pair).

72. The geometric structure of $CH_4$ is that of a tetrahedron. The carbon atom of $CH_4$ is surrounded by four bonding electron pairs. The H–C–H bond angle is the characteristic angle of the tetrahedron, 109.5°.

## 12.9 Molecular Structure: The VSEPR Model

74. The general molecular structure of a molecule is determined by how many electron pairs surround the central atom in the molecule, and by which of those electron pairs are used for bonding to the other atoms of the molecule.

76. Beryllium atoms only have *two* valence electrons. In $BeF_2$, there are single bonds between the beryllium atom and each fluorine atom. The beryllium atom of $BeF_2$ thus has two electron pairs around it, which lie 180° apart from one another. For the water molecule, in addition to the bonding pairs of electrons which attach the hydrogen atoms to the oxygen atoms, there are two nonbonding pairs of electrons which affect the H–O–H bond angle.

78. In $NF_3$, the nitrogen atom has *four* pairs of valence electrons, whereas in $BF_3$, there are only *three* pairs of valence electrons around the boron atom. The nonbonding pair on nitrogen in $NF_3$ pushes the three F atoms out of the plane of the N atom.

80. a. four electron pairs arranged tetrahedrally about C

    b. four electron pairs arranged tetrahedrally about Ge

    c. three electron pairs arranged trigonally (planar) around B

82. a. nonlinear (*V*-shaped, due to lone pairs on O)

    b. nonlinear (*V*-shaped, due to lone pairs on O)

    c. tetrahedral

84. a. $ClO_3^-$, trigonal pyramid (lone pair on Cl)

b. $ClO_2^-$, nonlinear (*V*-shaped, two lone pairs on Cl)

c. $ClO_4^-$, tetrahedral (all pairs on Cl are bonding)

86. a. 180°

b. 120°

c. <109.5°

d. 109.5°

## Additional Problems

88. In a covalent bond between two atoms of the same element, the electron pair is shared equally and the bond is nonpolar; with a bond between atoms of different elements, the electron pair is unequally shared and the bond is polar (assuming the elements have different electronegativities).

90. double

92. The bond with the larger electronegativity difference will be the more polar bond. See Figure 12.3 for electronegativities.

a. S–F

b. P–O

c. C–H

94. a. $BeF_2$

$$:\ddot{\underset{..}{F}}\text{—Be—}\ddot{\underset{..}{F}}:$$

b. $BF_3$

$$:\ddot{\underset{..}{F}}\text{—}\underset{\underset{:\underset{..}{F}:}{|}}{B}\text{—}\ddot{\underset{..}{F}}:$$

c. $NO_2$

$$:\ddot{O}=\ddot{N}\text{—}\ddot{\underset{..}{O}}\cdot \leftrightarrow \cdot\ddot{\underset{..}{O}}\text{—}\ddot{N}=\ddot{O} \leftrightarrow$$

$$:\ddot{O}=\dot{N}\text{—}\ddot{\underset{..}{O}}: \leftrightarrow :\ddot{\underset{..}{O}}\text{—}\dot{N}=\ddot{O}:$$

d. NO

$$\cdot\ddot{N}=\ddot{O}: \leftrightarrow :\ddot{N}=\ddot{O}\cdot$$

96. a. $SbCl_3$; trigonal pyramid (lone pair on Sb)

```
: Cl—Sb—Cl :
      |
    : Cl :
```

b. $GeF_4$; tetrahedral (all pairs on Ge are bonding)

```
     : F :
       |
: F—Ge—F :
       |
     : F :
```

c. $SF_2$; nonlinear (*V*-shaped, two lone pairs on S)

```
: F—S—F :
```

d. $SeCl_2$; nonlinear (*V*-shaped, two lone pairs on Se)

```
: Cl—Se—Cl :
```

## Chapter Thirteen Gases

### 13.1 Pressure

2. The "pressure of the atmosphere" represents the weight of the several-mile-thick layer of gases pressing down on every surface of the earth. Atmospheric pressure is most commonly measured with a mercury barometer (the pressure of the atmosphere is sufficient to maintain the height of a column of mercury to approximately 76 cm at sea level). The pressure of the atmosphere varies on the surface of the earth due to weather conditions, and also varies with altitude.

4. 760

6. less

8. a. $$665 \text{ mm Hg} \times \frac{1 \text{ atm}}{760 \text{ mm Hg}} = 0.875 \text{ atm}$$

b. $124 \text{ kPa} = 124 \times 10^3 \text{ Pa}$

$$124 \times 10^3 \text{ Pa} \times \frac{1 \text{ atm}}{101{,}325 \text{ Pa}} = 1.22 \text{ atm}$$

c. $$2.540 \times 10^6 \text{ Pa} \times \frac{1 \text{ atm}}{101{,}325 \text{ Pa}} = 25.07 \text{ atm}$$

d. $$803 \text{ torr} \times \frac{1 \text{ atm}}{760 \text{ torr}} = 1.06 \text{ atm}$$

10. a. $$0.903 \text{ atm} \times \frac{760 \text{ mm Hg}}{1 \text{ atm}} = 686 \text{ mm Hg}$$

b. $$2.1240 \times 10^6 \text{ Pa} \times \frac{1 \text{ atm}}{101{,}325 \text{ Pa}} \times \frac{760 \text{ mm Hg}}{1 \text{ atm}} = 1.5931 \times 10^4 \text{ mm Hg}$$

c. $445 \text{ kPa} = 445 \times 10^3 \text{ Pa}$

$$445 \times 10^3 \text{ Pa} \times \frac{1 \text{ atm}}{101{,}325 \text{ Pa}} \times \frac{760 \text{ mm Hg}}{1 \text{ atm}} = 3.34 \times 10^3 \text{ mm Hg}$$

d. $342 \text{ torr} = 342 \text{ mm Hg}$

12. a. $$645 \text{ mm Hg} \times \frac{1 \text{ atm}}{760 \text{ mm Hg}} \times \frac{101{,}325 \text{ Pa}}{1 \text{ atm}} = 8.60 \times 10^4 \text{ Pa}$$

b. $221 \text{ kPa} = 221 \times 10^3 \text{ Pa} = 2.21 \times 10^5 \text{ Pa}$

c. $$0.876 \text{ atm} \times \frac{101{,}325 \text{ Pa}}{1 \text{ atm}} = 8.87 \times 10^4 \text{ Pa}$$

d. $$32 \text{ torr} \times \frac{1 \text{ atm}}{760 \text{ torr}} \times \frac{101{,}325 \text{ Pa}}{1 \text{ atm}} = 4.3 \times 10^3 \text{ Pa}$$

## 13.2 Pressure and Volume: Boyle's Law

14. increases

16. $PV = k$; $P_1V_1 = P_2V_2$

18. a. 1002 mm Hg = 1.318 atm

$$V = 123 \text{ L} \times \frac{4.56 \text{ atm}}{1.318 \text{ atm}} = 425 \text{ L}$$

b. 25.2 mm Hg = 0.0332 atm

$$V = 0.0332 \text{ atm} \times \frac{634 \text{ mL}}{166 \text{ mL}} = 0.127 \text{ atm}$$

c. 511 torr = $6.81 \times 10^4$ Pa = 68.1 kPa

$$V = 44.3 \text{ L} \times \frac{68.1 \text{ kPa}}{1.05 \text{ kPa}} = 2.87 \times 10^3 \text{ L}$$

20. a. 1.00 mm Hg = 1.00 torr

$$V = 255 \text{ mL} \times \frac{1.00 \text{ torr}}{2.00 \text{ torr}} = 128 \text{ mL}$$

b. 1.0 atm = 101.325 kPa

$$V = 1.3 \text{ L} \times \frac{1.0 \text{ kPa}}{101.325 \text{ kPa}} = 1.3 \times 10^{-2} \text{ L}$$

c. 1.0 mm Hg = 0.133 kPa

$$V = 1.3 \text{ L} \times \frac{1.0 \text{ kPa}}{0.133 \text{ kPa}} = 9.8 \text{ L}$$

22. Assume the pressure at sea level to be 1 atm (760 mm Hg). Calculate the volume the balloon would have if it rose to the point where the pressure has dropped to 500 mm Hg. If this calculated volume is greater than the balloon's specific maximum volume (2.5 L) the balloon will burst.

$$2.0\ \text{L} \times \frac{760\ \text{mm Hg}}{500\ \text{mm Hg}} = 3.0\ \text{L} > 2.5\ \text{L}.$$ The balloon will burst.

24. 1.105 L = 1105 mL

$$755\ \text{torr} \times \frac{1105\ \text{mL}}{1.00\ \text{mL}} = 8.34 \times 10^5\ \text{torr}\ (1.10 \times 10^3\ \text{atm})$$

## 13.3 Volume and Temperature: Charles's Law

26. absolute zero (0 K)

28. $V = bT$; $V_1/T_1 = V_2/T_2$

30. 22°C + 273 = 295 K 100°C + 273 = 373 K

$$729\ \text{mL} \times \frac{373\ \text{K}}{295\ \text{K}} = 922\ \text{mL}$$

32. a. 75°C + 273 = 348 K −75°C + 273 = 198 K

$$100.\ \text{mL} \times \frac{198\ \text{K}}{348\ \text{K}} = 56.9\ \text{mL}$$

b. 100°C + 273 = 373 K

$$373\ \text{K} \times \frac{600\ \text{mL}}{500\ \text{mL}} = 448\ \text{K}\ (175°\text{C})$$

c. zero (the volume of any gas sample becomes zero at 0 K)

34. a. 0°C + 273 = 273 K

$$273\ \text{K} \times \frac{44.4\ \text{L}}{22.4\ \text{L}} = 541\ \text{K}\ (268°\text{C})$$

b. −272°C + 273 = 1 K 25°C + 273 = 298 K

$$1.0 \times 10^{-3}\ \text{mL} \times \frac{298\ \text{K}}{1\ \text{K}} = 0.30\ \text{mL}$$

c. $-40°C + 273 = 233\ K$

$$233\ K \times \frac{1000\ L}{32.3\ L} = 7.21 \times 10^3\ K\ (6940°C)$$

36. $25°C + 273 = 298\ K$ $\qquad -100°C + 273 = 173\ K$

$$25\ L \times \frac{173\ K}{298\ K} = 15\ L$$

38. One method of solution (using Charles's Law in the form $V_1/T_1 = V_2/T_2$) is shown in Example 13.6. A second method might make use of Charles's Law in the form $V = kT$. Using the information that the gas thermometer has a volume of 135 mL at 11°C (284 K), we can solve for the value of the proportionality constant $k$ in the formula, and then use this information to calculate the additional temperatures requested.

$V = kT$

$$k = \frac{V}{T} = \frac{135\ mL}{284\ K} = 0.475\ mL/K$$

$$T = \frac{V}{k} = \frac{V}{0.475\ mL/K} = 2.11V$$

For 113 mL, T = 2.11(113) = 238 K (−35°C)

For 142 mL, T = 2.11(142) = 299 K (26°C)

For 155 mL, T = 2.11(155) = 326 K (53°C)

For 127 mL, T = 2.11(127) = 267 K (−6°C)

## 13.4 Volume and Moles: Avogadro's Law

40. $V = an;$ $\qquad V_1/n_1 = V_2/n_2$

42. 4.0 g He = 1.0 mol $\qquad$ 3.0 g He = 0.75 mol

$$22.4\ L \times \frac{0.75\ mol}{1.0\ mol} = 17\ L$$

Note that, since the number of moles of Ne (or any gas) present in a sample is *directly proportional* to the mass of the gas sample, the problem could also have been set up directly in terms of the masses.

$$22.4\ L \times \frac{3.0\ g\ He}{4.0\ g\ He} = 17\ L$$

44. For a given gas, the number of moles present in a sample is directly proportional to the mass of the sample. So the problem can be solved even though the gas is not identified (so that its molar mass is not known).

$$23.2 \text{ g} \times \frac{10.4 \text{ L}}{93.2 \text{ L}} = 2.59 \text{ g}$$

## 13.5 The Ideal Gas Law

46. To give a numerical value of 0.08206 for the universal gas constant, the volume of the gas must be expressed in liters, the pressure in atmospheres, the amount of gas in moles, and the temperature of the gas in Kelvins.

$$R = 0.08206 \frac{\text{L atm}}{\text{mol K}} = 0.08206 \text{ L atm mol}^{-1} \text{ K}^{-1}$$

48. Boyle's Law: constant amount of gas and constant temperature.

$$PV = nRT$$

$$PV = (\text{constant})R(\text{constant})$$

$$PV = \text{constant}$$

Charles's Law: constant pressure and constant amount of gas.

$$PV = nRT.$$

$$(\text{constant})V = (\text{constant})RT$$

$$V = (\text{constant})T$$

Avogadro's Law: constant pressure and temperature.

$$PV = nRT$$

$$(\text{constant})V = nR(\text{constant})$$

$$V = (\text{constant})n$$

50. a. $V = 142 \text{ mL} = 0.142 \text{ L}$

$$T = PV/nR = \frac{(21.2 \text{ atm})(0.142 \text{ L})}{(0.432 \text{ mol})(0.08206 \text{ L atm mol}^{-1} \text{ K}^{-1})} = 84.9 \text{ K}$$

b. $V = 1.23 \text{ mL} = 0.00123 \text{ L}$

$$P = nRT/V = \frac{(0.000115 \text{ mol})((0.08206 \text{ L atm mol}^{-1} \text{ K}^{-1})(293 \text{ K})}{(0.00123 \text{ L})}$$

$P = 2.25 \text{ atm}$

c. $P = 755 \text{ mm Hg} = 0.993 \text{ atm}$

$T = 131°\text{C} + 273 = 404 \text{ K}$

$$V = nRT/P = \frac{(0.473 \text{ mol})(0.08206 \text{ L atm mol}^{-1} \text{ K}^{-1})(404 \text{ K})}{(0.993 \text{ atm})} = 15.8 \text{ L}$$

$15.8 \text{ L} = 1.58 \times 10^4 \text{ mL}$

52. a. $V = 21.2 \text{ mL} = 0.0212 \text{ L}$

$$T = PV/nR = \frac{(1.034 \text{ atm})(0.0212 \text{ L})}{(0.00432 \text{ mol})(0.08206 \text{ L atm mol}^{-1} \text{ K}^{-1})} = 61.8 \text{ K}$$

b. $V = 1.73 \text{ mL} = 0.00173 \text{ L}$

$$P = nRT/V = \frac{(0.000115 \text{ mol})(0.08206 \text{ L atm mol}^{-1} \text{ K}^{-1})(182 \text{ K})}{(0.00173 \text{ L})}$$

$P = 0.993 \text{ atm}$

c. $P = 1.23 \text{ mm Hg} = 0.00162 \text{ atm}$

$T = 152°\text{C} + 273 = 425 \text{ K}$

$$V = nRT/P = \frac{(0.773 \text{ mol})(0.08206 \text{ L atm mol}^{-1} \text{ K}^{-1})(425 \text{ K})}{(0.00162 \text{ atm})}$$

$V = 1.67 \times 10^4 \text{ L}$

54. molar mass $SO_2$ = 64.06 g

$$n = 32.0 \text{ g } SO_2 \times \frac{1 \text{ mol}}{64.06 \text{ g } SO_2} = 0.500 \text{ mol}$$

$T = 22°\text{C} + 273 = 295 \text{ K}$

$$V = nRT/P = \frac{(0.500 \text{ mol})(0.08206 \text{ L atm mol}^{-1} \text{ K}^{-1})(295 \text{ K})}{(1.054 \text{ atm})} = 11.5 \text{ L}$$

56. $27°C + 273 = 300\ K$

The number of moles of gas it takes to fill the 100. L tanks to 120 atm at 27°C is independent of the identity of the gas.

$$n = PV/RT = \frac{(120\ atm)(100.\ L)}{(0.08206\ L\ atm\ mol^{-1}\ K^{-1})(300\ K)} = 487\ mol$$

487 mol of *any* gas will fill the tanks to the required specifications.

molar masses: $CH_4$, 16.0 g; $N_2$, 28.0 g; $CO_2$, 44.0 g

for $CH_4$: (487 mol)(16.0 g/mol) = 7792 g = 7.79 kg $CH_4$

for $N_2$: (487 mol)(28.0 g/mol) = 13,636 g = 13.6 kg $N_2$

for $CO_2$: (487 mol)(44.0 g/mol) = 21,428 g = 21.4 kg $CO_2$

58. molar mass He = 4.003 g

$$n = 4.00\ g\ He \times \frac{1\ mol}{4.003\ g\ He} = 0.999\ mol$$

$$T = PV/nR = \frac{(1.00\ atm)(22.4\ L)}{(0.999\ mol)(0.08206\ L\ atm\ mol^{-1}\ K^{-1})} = 273\ K\ (0°C)$$

60. molar mass of $O_2$ = 32.00 g

55 mg = 0.055 g

$$n = 0.055\ g \times \frac{1\ mol}{32.00\ g} = 0.0017\ mol$$

$V$ = 100. mL = 0.100 L

$T$ = 26°C + 273 = 299 K

$$P = nRT/V = \frac{(0.0017\ mol)(0.08206\ L\ atm\ mol^{-1}\ K^{-1})(299\ K)}{(0.100\ L)} = 0.42\ atm$$

62. $P_1$ = 1.0 atm $\quad$ $P_2$ = 220 torr = 0.289 atm

$V_1$ = 1.0 L $\quad$ $V_2$ = ?

$T_1$ = 23°C + 273 = 296 K $\quad$ $T_2$ = −31°C = 242 K

$$V_2 = \frac{T_2P_1V_1}{T_1P_2} = \frac{(242\ K)(1.0\ atm)(1.0\ L)}{(296\ K)(0.289\ atm)} = 2.8\ L$$

64. $P_1 = 1.13$ atm $\quad P_2 = 1.89$ atm

$V_1 = 100$ mL $= 0.100$ L $\quad V_2 = 500$ mL $= 0.500$ L

$T_1 = 300$ K $\quad T_2 = ?$

$$T_2 = \frac{T_1 P_2 V_2}{P_1 V_1} = \frac{(300\ \text{K})(1.89\ \text{atm})(0.500\ \text{L})}{(1.13\ \text{atm})(0.100\ \text{L})} = 2.51 \times 10^3\ \text{K}$$

Note that the calculation could have been carried through with the two volumes expressed in milliliters since the universal gas constant does not appear explicitly in this form of the ideal gas equation.

## 13.6 Dalton's Law of Partial Pressures

66. The total pressure in a sample of gas which has been collected by bubbling through water is made up of two components: the pressure of the gas of interest and the pressure of water vapor. The partial pressure of the gas of interest is then the total pressure of the sample minus the vapor pressure of water.

68. molar mass of $O_2$ = 32.00 g

$$50.\ \text{g}\ O_2 \times \frac{1\ \text{mol}\ O_2}{32.00\ \text{g}\ O_2} = 1.56\ \text{mol}\ O_2$$

total number of moles of gas = 1.0 mol $N_2$ + 1.56 mol $O_2$ = 2.56 mol

25°C + 273 = 298 K

$$P = nRT/V = \frac{(2.56\ \text{mol})(0.08206\ \text{L atm mol}^{-1}\ \text{K}^{-1})(298\ \text{K})}{(5.0\ \text{L})} = 13\ \text{atm}$$

70. molar masses: $O_2$, 32.0 g; $N_2$, 28.0 g; $CO_2$, 44.0 g; Ne, 20.2 g

$$5.0\ \text{g}\ O_2 \times \frac{1\ \text{mol}\ O_2}{32.0\ \text{g}\ O_2} = 0.156\ \text{mol}\ O_2$$

$$5.0\ \text{g}\ N_2 \times \frac{1\ \text{mol}\ N_2}{28.0\ \text{g}\ N_2} = 0.179\ \text{mol}\ N_2$$

$$5.0\ \text{g}\ CO_2 \times \frac{1\ \text{mol}\ CO_2}{44.0\ \text{g}\ CO_2} = 0.114\ \text{mol}\ CO_2$$

$$5.0\ \text{g Ne} \times \frac{1\ \text{mol Ne}}{20.2\ \text{g Ne}} = 0.248\ \text{mol Ne}$$

Total moles of gas = 0.156 + 0.179 + 0.114 + 0.248 = 0.697 mol

29°C + 273 = 302 K

755 mm Hg = 0.993 atm

$$V = nRT/P = \frac{(0.697\ \text{mol})(0.08206\ \text{L atm mol}^{-1}\ \text{K}^{-1})(302\ \text{K})}{(0.993\ \text{atm})} = 17\ \text{L}$$

72. The pressures must be expressed in the same units, either mm Hg or atm.

$P_{\text{hydrogen}} = P_{\text{total}} - P_{\text{water vapor}}$

1.023 atm = 777.5 mm Hg

$P_{\text{hydrogen}}$ = 777.5 mm Hg − 42.2 mm Hg = 735.3 mm Hg

42.2 mm Hg = 0.056 atm

$P_{\text{hydrogen}}$ = 1.023 atm − 0.056 atm = 0.967 atm

74. 1.032 atm = 784.3 mm Hg

$P_{\text{hydrogen}}$ = 784.3 mm Hg − 32 mm Hg = 752.3 mm Hg = 0.990 atm

240 mL = 0.240 L

30°C + 273 = 303 K

$$n_{\text{hydrogen}} = P_{\text{hydrogen}}V/RT = \frac{(0.990 \text{ atm})(0.240 \text{ L})}{(0.08206 \text{ L atm mol}^{-1} \text{ K}^{-1})(303 \text{ K})} = 0.00956 \text{ mol}$$

$$0.00956 \text{ mol } H_2 \times \frac{1 \text{ mol Zn}}{1 \text{ mol } H_2} = 0.00956 \text{ mol of Zn must have reacted}$$

molar mass of Zn = 65.38 g

$$0.00956 \text{ mol Zn} \times \frac{65.38 \text{ g Zn}}{1 \text{ mol Zn}} = 0.625 \text{ g Zn must have reacted}$$

## 13.7 Laws and Models: A Review

76. A theory is successful if it explains known experimental observations. Theories which have been successful in the past may not be successful in the future (for example, as technology evolves, more sophisticated experiments may be possible in the future).

## 13.8 The Kinetic Molecular Theory of Gases

78. collisions

80. no

## 13.9 The Implications of the Kinetic Molecular Theory

82. If the temperature of a sample of gas is increased, the average kinetic energy of the particles of gas increases. This means that the speeds of the particles increase. If the particles have a higher speed, they will hit the walls of the container more frequently and with greater force, thereby increasing the pressure.

## 13.10 Gas Stoichiometry

84. STP = 0°C, 1 atm pressure. These conditions were chosen because they are easy to attain and reproduce *experimentally*. The barometric pressure within a laboratory is likely to be near 1 atm most days, and 0°C can be attained with a simple ice bath.

86. molar mass of $NH_3$ = 17.03 g

$$5.00 \text{ g } NH_3 \times \frac{1 \text{ mol } NH_3}{17.03 \text{ g } NH_3} = 0.294 \text{ mol } NH_3 \text{ to be produced}$$

$$N_2(g) + 3H_2(g) \rightarrow 2NH_3(g)$$

$$0.294 \text{ mol } NH_3 \times \frac{1 \text{ mol } N_2}{2 \text{ mol } NH_3} = 0.147 \text{ mol } N_2 \text{ required}$$

$$0.294 \text{ mol } NH_3 \times \frac{3 \text{ mol } H_2}{2 \text{ mol } NH_3} = 0.441 \text{ mol } H_2 \text{ required}$$

11°C + 273 = 284 K

$$V_{\text{nitrogen}} = \frac{(0.147 \text{ mol})(0.08206 \text{ L atm mol}^{-1} \text{ K}^{-1})(284 \text{ K})}{(0.998 \text{ atm})} = 3.43 \text{ L } N_2$$

$$V_{\text{hydrogen}} = \frac{(0.441 \text{ mol})(0.08206 \text{ L atm mol}^{-1} \text{ K}^{-1})(284 \text{ K})}{(0.998 \text{ atm})} = 10.3 \text{ L } H_2$$

88. $C_6H_{12}O_6(s) + 6O_2(g) \rightarrow 6CO_2(g) + 6H_2O(g)$

molar mass of $C_6H_{12}O_6$ = 180. g

$$5.00 \text{ g } C_6H_{12}O_6 \times \frac{1 \text{ mol } C_6H_{12}O_6}{180. \text{ g } C_6H_{12}O_6} = 0.02777 \text{ mol } C_6H_{12}O_6$$

$$0.02777 \text{ mol } C_6H_{12}O_6 \times \frac{6 \text{ mol } O_2}{1 \text{ mol } C_6H_{12}O_6} = 0.1666 \text{ mol } O_2$$

28°C + 273 = 301 K

$$V_{\text{oxygen}} = \frac{(0.1666 \text{ mol})(0.08206 \text{ L atm mol}^{-1} \text{ K}^{-1})(301 \text{ K})}{(0.976 \text{ atm})} = 4.21 \text{ L}$$

Because the coefficients of $CO_2(g)$ and $H_2O(g)$ in the balanced chemical equation happen to be the same as the coefficient of $O_2(g)$, the calculations for the volumes of these gases produced are identical: 4.21 L of each gaseous product is produced.

90. $2Cu_2S(s) + 3O_2(g) \rightarrow 2Cu_2O(s) + 2SO_2(g)$

molar mass $Cu_2S$ = 159.2 g

$$25 \text{ g } Cu_2S \times \frac{1 \text{ mol } Cu_2S}{159.2 \text{ g } Cu_2S} = 0.1570 \text{ mol } Cu_2S$$

$$0.1570 \text{ mol } Cu_2S \times \frac{3 \text{ mol } O_2}{2 \text{ mol } Cu_2S} = 0.2355 \text{ mol } O_2$$

27.5°C + 273 = 300.5 K

$$V_{\text{oxygen}} = \frac{(0.2355 \text{ mol})(0.08206 \text{ L atm mol}^{-1} \text{ K}^{-1})(300.5 \text{ K})}{(0.998 \text{ atm})} = 5.8 \text{ L } O_2$$

$$0.1570 \text{ mol } Cu_2S \times \frac{2 \text{ mol } SO_2}{2 \text{ mol } Cu_2S} = 0.1570 \text{ mol } SO_2$$

$$V_{\text{sulfur dioxide}} = \frac{(0.1570 \text{ mol})(0.08206 \text{ L atm mol}^{-1} \text{ K}^{-1})(300.5 \text{ K})}{(0.998 \text{ atm})}$$

$$V_{\text{sulfur dioxide}} = 3.9 \text{ L } SO_2$$

92. $2NaHCO_3(s) \rightarrow Na_2CO_3(s) + H_2O(g) + CO_2(g)$

molar mass $NaHCO_3$ = 84.01 g

$$1.00 \text{ g } NaHCO_3 \times \frac{1 \text{ mol } NaHCO_3}{84.01 \text{ g } NaHCO_3} = 0.01190 \text{ mol } NaHCO_3$$

$$0.01190 \text{ mol } NaHCO_3 \times \frac{1 \text{ mol } H_2O}{2 \text{ mol } NaHCO_3} = 0.00595 \text{ mol } H_2O$$

Because $H_2O(g)$ and $CO_2(g)$ have the same coefficients in the balanced chemical equation for the reaction, if 0.00595 mol $H_2O$ is produced, then 0.00595 mol $CO_2$ must also be produced. The total number of moles of gaseous substances produced is thus 0.00595 + 0.00595 = 0.0119 mol.

29°C + 273 = 302 K

769 torr = 1.012 atm

$$V_{\text{total}} = \frac{(0.0119 \text{ mol})(0.08206 \text{ L atm mol}^{-1} \text{ K}^{-1})(302 \text{ K})}{(1.012 \text{ atm})} = 0.291 \text{ L}$$

94. One mole of any ideal gas occupies 22.4 L at STP.

$$35 \text{ mol } N_2 \times \frac{22.4 \text{ L}}{1 \text{ mol}} = 7.8 \times 10^2 \text{ L}$$

96. $P_1 = 0.987$ atm $\quad P_2 = 1.00$ atm

$V_1 = 125$ L $\quad V_2 = ?$

$T_1 = 25°C + 273 = 298$ K $\quad T_2 = 273$ K

$$V_2 = \frac{T_2P_1V_1}{T_1P_2} = \frac{(273 \text{ K})(0.987 \text{ atm})(125 \text{ L})}{(298 \text{ K})(1.00 \text{ atm})} = 113 \text{ L}$$

98. molar masses: He, 4.003 g; Ar, 39.95 g; Ne, 20.18 g

$$5.0 \text{ g He} \times \frac{1 \text{ mol He}}{4.003 \text{ g He}} = 1.249 \text{ mol He}$$

$$1.0 \text{ g Ar} \times \frac{1 \text{ mol Ar}}{39.95 \text{ g Ar}} = 0.02503 \text{ mol Ar}$$

$$3.5 \text{ g Ne} \times \frac{1 \text{ mol Ne}}{20.18 \text{ g Ne}} = 0.1734 \text{ mol Ne}$$

Total moles of gas = 1.249 + 0.02503 + 0.1734 = 1.447 mol

22.4 L is the volume occupied by one mole of any ideal gas at STP. This would apply even if the gas sample is a *mixture* of individual gases.

$$1.447 \text{ mol} \times \frac{22.4 \text{ L}}{1 \text{ mol}} = 32 \text{ L total volume for the mixture}$$

The *partial pressure* of each individual gas in the mixture will be related to what *fraction* on a mole basis each gas represents in the mixture.

$$P_{He} = 1.00 \text{ atm} \times \frac{1.249 \text{ mol He}}{1.447 \text{ mol total}} = 0.86 \text{ atm}$$

$$P_{Ar} = 1.00 \text{ atm} \times \frac{0.02503 \text{ mol Ar}}{1.447 \text{ mol total}} = 0.017 \text{ atm}$$

$$P_{Ne} = 1.00 \text{ atm} \times \frac{0.1734 \text{ mol Ne}}{1.447 \text{ mol total}} = 0.12 \text{ atm}$$

100. $N_2(g) + 3H_2(g) \rightarrow 2NH_3(g)$

molar mass of $N_2$ = 28.02 g

$$10.0 \text{ g } N_2 \times \frac{1 \text{ mol } N_2}{28.02 \text{ g } N_2} = 0.3569 \text{ mol } N_2$$

$$0.3569 \text{ mol } N_2 \times \frac{3 \text{ mol } H_2}{1 \text{ mol } N_2} = 1.071 \text{ mol } H_2$$

$$1.071 \text{ mol } H_2 \times \frac{22.4 \text{ L}}{1 \text{ mol}} = 24.0 \text{ L}$$

102. $2K_2MnO_4(aq) + Cl_2(g) \rightarrow 2KMnO_4(s) + 2KCl(aq)$

molar mass $KMnO_4$ = 158.0 g

$$10.0 \text{ g } KMnO_4 \times \frac{1 \text{ mol } KMnO_4}{158.0 \text{ g } KMnO_4} = 0.06329 \text{ mol } KMnO_4$$

$$0.06329 \text{ mol } KMnO_4 \times \frac{1 \text{ mol } Cl_2}{2 \text{ mol } KMnO_4} = 0.03165 \text{ mol } Cl_2$$

$$0.03165 \text{ mol } Cl_2 \times \frac{22.4 \text{ L}}{1 \text{ mol}} = 0.708 \text{ L} = 708 \text{ mL}$$

## Additional Problems

104. twice

106. a. $PV = k;\ P_1V_1 = P_2V_2$

b. $V = bT;\ V_1/T_1 = V_2/T_2$

c. $V = an;\ V_1/n_1 = V_2/n_2$

d. $PV = nRT$

e. $P_1V_1/T_1 = P_2V_2/T_2$

108. First determine what volume the helium in the tank would have if it were at a pressure of 755 mm Hg (corresponding to the pressure the gas will have in the balloons).

8.40 atm = 6384 mm Hg

$$V_2 = (25.2 \text{ L}) \times \frac{6384 \text{ mm Hg}}{755 \text{ mm Hg}} = 213 \text{ L}$$

Allowing for the fact that 25.2 L of He will have to remain in the tank, this leaves 213 − 25.2 = 187.8 L of He for filling the balloons.

$$187.8 \text{ L He} \times \frac{1 \text{ balloon}}{1.5 \text{ L He}} = 125 \text{ balloons}$$

110. The volume of a gas sample is proportional to the number of moles of gas present in the sample (Avogadro's law). Therefore, the volumes of gases combining in chemical reactions should follow the stoichiometric ratios of the balanced chemical equation for the reaction, assuming the gases involved are all subject to the same conditions of temperature and pressure.

For the reaction $2NO_2(g) \rightarrow N_2O_4(g)$, we ordinarily would interpret the equation as meaning that two *moles* of $NO_2$ will combine to produce one *mole* of $N_2O_4$. However, if the $NO_2$ and $N_2O_4$ are each at the same pressure and temperature, we could just as easily interpret the equation to mean that two *volumes* of $NO_2$ combine to give one *volume* of $N_2O_4$.

$$25.0 \text{ mL } NO_2 \times \frac{1 \text{ mL } N_2O_4}{2 \text{ mL } NO_2} = 12.5 \text{ mL } N_2O_4$$

112. $CaCO_3(s) + 2H^+(aq) \rightarrow Ca^{2+}(aq) + H_2O(l) + CO_2(g)$

molar mass $CaCO_3$ = 100.1 g

$$10.0 \text{ g } CaCO_3 \times \frac{1 \text{ mol } CaCO_3}{100.1 \text{ g } CaCO_3} = 0.0999 \text{ mol } CaCO_3$$

60°C + 273 = 333 K

$$P_{\text{carbon dioxide}} = P_{\text{total}} - P_{\text{water vapor}}$$

$$P_{\text{carbon dioxide}} = 774 \text{ mm Hg} - 149.4 \text{ mm Hg} = 624.6 \text{ mm Hg} = 0.822 \text{ atm}$$

$$V_{\text{wet}} = \frac{(0.0999 \text{ mol})(0.08206 \text{ L atm mol}^{-1} \text{ K}^{-1})(333 \text{ K})}{(0.822 \text{ atm})} = 3.32 \text{ L wet } CO_2$$

$$V_{\text{dry}} = 3.32 \text{ L} \times \frac{624.6 \text{ mm Hg}}{774 \text{ mm Hg}} = 2.68 \text{ L}$$

114. $2KClO_3(s) \rightarrow 2KCl(s) + 3O_2(g)$

molar mass $KClO_3$ = 122.6 g

$$50.0 \text{ g } KClO_3 \times \frac{1 \text{ mol } KClO_3}{122.6 \text{ g } KClO_3} = 0.408 \text{ mol } KClO_3$$

$$0.408 \text{ mol } KClO_3 \times \frac{3 \text{ mol } O_2}{2 \text{ mol } KClO_3} = 0.612 \text{ mol } O_2$$

25°C + 273 = 298 K

630. torr = 0.829 atm

$$V = nRT/P = \frac{(0.612)(0.08206 \text{ L atm mol}^{-1} \text{ K}^{-1})(298 \text{ K})}{(0.829 \text{ atm})} = 18.1 \text{ L } O_2$$

# Chapter Fourteen Liquids and Solids

## 14. Water and Its Phase Changes

2. Water is the solvent in which cellular processes take place in living creatures. Water in the oceans moderates the earth's temperature. Water is used in industry as a cooling agent. Water serves as a means of transportation on the earth's oceans. The liquid range is 0°C to 100°C.

4. The fact that water expands when it freezes often results in broken water pipes during cold weather. The expansion of water when it freezes also makes ice float on liquid water.

6. From room temperature to the freezing point (0°C), the average kinetic energy of the molecules in liquid water decreases, and the molecules slow down. At the freezing point, the liquid freezes, with the molecules forming a crystal lattice in which there is much greater order than in the liquid state: molecules no longer move freely, but rather are only able to vibrate somewhat. Below the freezing point, the molecules' vibrations slow down as the temperature is lowered further.

## 14.2 Energy Requirements for the Changes of State

8. At higher altitudes, the boiling points of liquids, such as water, are lower because there is a lower atmospheric pressure above the liquid. The temperature at which food cooks is determined by the temperature to which the water in the food can be heated before it escapes as steam. Thus, food cooks at a lower temperature at high elevations where the boiling point of water is lowered.

10. *Intra*molecular forces are the forces that hold the atoms together within a molecule. When a molecular solid is melted, it is the forces between molecules, not within the molecules themselves, that must be overcome. *Intra*molecular forces are typically stronger than *inter*molecular forces.

12. Heat of fusion (melt); Heat of vaporization (boil). The heat of vaporization is always larger, because virtually all of the intermolecular forces must be overcome to form a gas. In a liquid, considerable intermolecular forces remain. Thus going from a solid to liquid requires less energy than going from the liquid to the gas.

14. $$10.0 \text{ g X} \times \frac{1 \text{ mol X}}{52 \text{ g X}} = 0.192 \text{ mol X}$$

$$0.192 \text{ mol} \times \frac{2.5 \text{ kJ}}{1 \text{ mol}} = 0.48 \text{ kJ}$$

$$25.0 \text{ g X} \times \frac{1 \text{ mol X}}{52 \text{ g X}} = 0.481 \text{ mol X}$$

$$0.481 \text{ mol} \times \frac{55.3 \text{ kJ}}{1 \text{ mol}} = 27 \text{ kJ}$$

16. molar mass $H_2O$ = 18.0 g

$$5.0 \text{ g } H_2O \times \frac{1 \text{ mol } H_2O}{18.0 \text{ g } H_2O} = 0.278 \text{ mol } H_2O$$

To melt the ice: $0.278 \text{ mol} \times \frac{6.02 \text{ kJ}}{1 \text{ mol}} = 1.67 \text{ kJ}$

To heat the liquid: $5.0 \text{ g} \times 4.18 \frac{\text{J}}{\text{g °C}} \times 100\text{°C} = 2090 \text{ J} = 2.09 \text{ kJ}$

To vaporize the liquid: $0.278 \text{ mol} \times \frac{40.6 \text{ kJ}}{1 \text{ mol}} = 11.27 \text{ kJ}$

Total heat required = 1.67 + 2.09 + 11.27 = 15.03 kJ = 15 kJ

18. molar mass $CS_2$ = 76.1 g

$$1.0 \text{ g } CS_2 \times \frac{1 \text{ mol } CS_2}{76.1 \text{ g } CS_2} = 0.0131 \text{ mol } CS_2$$

$$0.131 \text{ mol} \times \frac{28.4 \text{ kJ}}{1 \text{ mol}} = 0.37 \text{ kJ required}$$

$$50. \text{ g } CS_2 \times \frac{1 \text{ mol } CS_2}{76.1 \text{ g } CS_2} = 0.657 \text{ mol } CS_2$$

$$0.657 \text{ mol } CS_2 \times \frac{28.4 \text{ kJ}}{1 \text{ mol}} = 19 \text{ kJ evolved } (-19 \text{ kJ})$$

## 14.3 Intermolecular Forces

20. Dipole-dipole interactions are typically about 1% as strong as a covalent bond. Dipole-dipole interactions represent electrostatic attractions between portions of molecules which carry only a *partial* positive or negative charge, and such forces require the molecules that are interacting to come *near* enough to each other.

22. Hydrogen bonding is a particularly strong dipole-dipole force arising among molecules in which hydrogen is bound to a highly electronegative atom, such as N, O, or F. Such bonds are highly polar, and because the H atom is so small, the dipoles of different molecules are able to approach each other much more closely than in other polar molecules.

Examples of substances in which hydrogen bonding is important include $H_2O$, $NH_3$, and HF.

24. London dispersion forces are relatively weak forces that arise among noble gas atoms and in nonpolar molecules. London forces are due to *instantaneous dipoles* that develop when one atom (or molecule) momentarily distorts the electron cloud of another atom (or molecule). London forces are typically weaker than either permanent dipole-dipole forces or covalent bonds.

26. a. London dispersion forces (nonpolar molecules)

    b. hydrogen bonding (H attached to N)

    c. London dispersion forces (nonpolar molecules)

    d. London dispersion forces (nonpolar molecules)

28. An increase in the heat of fusion is observed for an increase in the size of the halogen atom involved (the electron cloud of a larger atom is more easily polarized by an approaching dipole, thus giving larger London dispersion forces).

30. For a homogeneous mixture to be able to form at all, the forces between molecules of the two substances being mixed must be at least *comparable in magnitude* to the intermolecular forces within each *separate* substance. Apparently in the case of a water-ethanol mixture, the forces that exist when water and ethanol are mixed are stronger than water-water or ethanol-ethanol forces in the separate substances. This allows ethanol and water molecules to approach each other more closely in the mixture than either substance's molecules could approach a like molecule in the separate substances. There is strong hydrogen bonding in both ethanol and water.

## 14.4 Evaporation and Vapor Pressure

32. Vapor pressure is the pressure of vapor present *at equilibrium* above a liquid in a sealed container at a particular temperature. When a liquid is placed in a closed container, molecules of the liquid evaporate freely into the empty space above the liquid. As the number of molecules present in the vapor state increases with time, vapor molecules begin to rejoin the liquid state (condense). Eventually a dynamic equilibrium is reached between evaporation and condensation in which the net number of molecules present in the vapor phase becomes *constant* with time.

34. A *volatile* liquid is one that evaporates relatively easily. Volatile liquids typically have large vapor pressures because the intermolecular forces that would tend to prevent evaporation are small.

36. a. HF — Although both substances are capable of hydrogen bonding, water has two O–H bonds which can be involved in hydrogen bonding versus only one F–H bond in HF.

    b. $CH_3OCH_3$ — Since there is no H attached to the O atom, no hydrogen bonding can exist. Since there is no hydrogen bonding possible, the molecule should be relatively more volatile than $CH_3CH_2OH$ even though it contains the same number of atoms of each element.

    c. $CH_3SH$ — Hydrogen bonding is not as important for a S–H bond (because S has a lower electronegativity than O). Since there is little hydrogen bonding, the molecule is relatively more volatile than $CH_3OH$.

38. In $NH_3$, strong hydrogen bonding can exist. In $CH_4$, because the molecule is nonpolar, only the relatively weak London dispersion forces exist.

## 14.5 The Solid State

40. *Ionic* solids have as their fundamental particles positive and negative *ions*; a simple example is sodium chloride, in which $Na^+$ and $Cl^-$ ions are held together by strong electrostatic forces.

    *Molecular* solids have molecules as their fundamental particles, with the molecules being held together in the crystal by dipole-dipole forces, hydrogen bonding forces, or London dispersion forces (depending on the identity of the substance); simple examples of molecular solids include ice ($H_2O$) and ordinary table sugar (sucrose).

    *Atomic* solids have simple atoms as their fundamental particles, with the atoms being held together in the crystal either by covalent bonding (as in graphite or diamond) or by metallic bonding (as in copper or other metals).

## 14.6 Bonding in Solids

42. The fundamental particles in molecular solids are individual *molecules*; ice (individual water molecules), sucrose (individual sucrose molecules)

44. Ionic solids typically have the highest melting points, because the ionic charges and close packing in ionic solids allow for very strong forces among a given ion and its nearest neighbors of the opposite charge.

46. In a crystal of ice, strong *hydrogen bonding* forces are present, while in the crystal of a nonpolar substance like oxygen, only the much weaker *London* forces exist.

48. The electron sea model envisions a metal as a cluster of positive ions through which the valence electrons are able to move freely. An electrical current represents the movement of electrons, for example

through a metal wire, and is consistent with a model in which the electrons are free to roam.

50. Alloys may be of two types: *substitutional* (in which one metal is substituted for another in the regular positions of the crystal lattice) and *interstitial* (in which a second metal's atoms fit into the empty space in a given metal's crystal lattice). The presence of atoms of a second metal in a given metal's crystal lattice changes the properties of the metal: frequently the alloy is stronger than either of the original metals because the irregularities introduced into the crystal lattice by the presence of a second metal's atoms prevent the crystal from being deformed as easily. The properties of iron may be modified by alloying with many different substances, particularly with carbon, nickel, and cobalt. Steels with relatively high carbon content are exceptionally strong, whereas steels with low carbon contents are softer, more malleable, and more ductile. Steels produced by alloying iron with nickel and cobalt are more resistant to corrosion than iron itself.

## Additional Problems

52. j

54. f

56. d

58. a

60. l

62. Dimethyl ether has the larger vapor pressure. No hydrogen bonding is possible since the O atom does not have a hydrogen atom attached. Hydrogen bonding can occur *only* when a hydrogen atom is *directly* attached to a strongly electronegative atom (such as N, O, or F). Hydrogen bonding *is* possible in ethanol (ethanol contains an –OH group).

64. a. $H_2$. London dispersion forces are the only intermolecular forces present in these nonpolar molecules; typically London forces become larger with increasing atomic size (as the atoms become bigger, the edge of the electron cloud lies farther from the nucleus and becomes more easily distorted).

    b. Xe. Only the relatively weak London forces could exist in a crystal of Xe atoms, whereas in NaCl strong ionic forces exist, and in diamond strong covalent bonding exists between carbon atoms.

    c. $Cl_2$. Only London forces exist among such nonpolar molecules. London forces become larger with increasing atomic size.

66. Steel is a general term applied to alloys consisting primarily of iron, but with small amounts of other substances added. Whereas pure iron itself is relatively soft, malleable, and ductile, steels are typically much stronger and harder, and much less subject to damage.

## Chapter Fifteen Solutions

### 15.1 Solubility

2. When an ionic solute dissolves in water, a given ion is pulled into solution by the attractive ion-dipole force exerted by several water molecules. For example, in dissolving a positive ion, the ion is approached by the negatively charged end of several water molecules: if the attraction of the water molecules for the positive ion is stronger than the attraction of the negative ions near it in the crystal, the ion leaves the crystal and enters solution. After entering solution, the dissolved ion is surrounded completely by water molecules, which tends to prevent the ion from reentering the crystal.

4. solid

6. In order for a molecular solid to dissolve in water, the forces between water molecules and solute molecules must be comparable to the forces in the crystal among the solute molecules themselves. Sugar molecules contain several –OH groups (making them similar in structure to water) which permit extensive hydrogen bonding between water and sugar molecules.

8. independently

### 15.2 Solution Composition: An Introduction

10. unsaturated

12. large

### 15.3 Solution Composition: Mass Percent

14. 0.1 g $CaCl_2$

16. a. $$\frac{5.0 \text{ g } KNO_3}{(5.0 \text{ g } KNO_3 + 75 \text{ g } H_2O)} \times 100 = \frac{5.0 \text{ g}}{80.0 \text{ g}} \times 100 = 6.3\% \; KNO_3$$

b. 2.5 mg = 0.0025 g

$$\frac{0.0025 \text{ g } KNO_3}{(0.025 \text{ g } KNO_3 + 1.0 \text{ g } H_2O)} \times 100 = \frac{0.0025 \text{ g}}{1.0025 \text{ g}} \times 100 = 0.25\% \; KNO_3$$

c. $$\frac{11 \text{ g } KNO_3}{(11 \text{ g } KNO_3 + 89 \text{ g } H_2O)} \times 100 = \frac{11 \text{ g}}{100 \text{ g}} \times 100 = 11\% \; KNO_3$$

d. $$\frac{11 \text{ g } KNO_3}{(11 \text{ g } KNO_3 + 49 \text{ g } H_2O)} \times 100 = \frac{11 \text{ g}}{60 \text{ g}} \times 100 = 18\% \; KNO_3$$

18. To say a solution is 25% by mass sugar means that 100 g of the solution would contain 25 g of sugar:

a. $$25 \text{ g sugar} \times \frac{100 \text{ g solution}}{25 \text{ g sugar}} = 100 \text{ g solution} = 1.0 \times 10^2 \text{ g}$$

b. $$5.0 \text{ g sugar} \times \frac{100 \text{ g solution}}{25 \text{ g sugar}} = 20. \text{ g solution}$$

c. $$100. \text{ g sugar} \times \frac{100 \text{ g solution}}{25 \text{ g sugar}} = 400 \text{ g solution} = 4.0 \times 10^2 \text{ g}$$

d. $1.0 \text{ kg} = 1.0 \times 10^3 \text{ g}$

$$1.0 \times 10^3 \text{ g} \times \frac{100 \text{ g solution}}{25 \text{ g sugar}} = 4.0 \text{ kg solution}$$

20. $$\%C = \frac{5.0 \text{ g C}}{(5.0 \text{ g C} + 1.5 \text{ g Ni} + 100. \text{ g Fe})} \times 100 = \frac{5.0 \text{ g}}{106.5 \text{ g}} \times 100 = 4.7\% \text{ C}$$

$$\%Ni = \frac{1.5 \text{ g Ni}}{(5.0 \text{ g C} + 1.5 \text{ g Ni} + 100. \text{ g Fe})} \times 100 = \frac{1.5 \text{ g}}{106.5 \text{ g}} \times 100 = 1.4\% \text{ Ni}$$

$$\%Fe = \frac{100. \text{ g Fe}}{(5.0 \text{ g C} + 1.5 \text{ g Ni} + 100. \text{ g Fe})} \times 100 = \frac{100. \text{ g}}{106.5 \text{ g}} \times 100 = 93.9\% \text{ Fe}$$

22. $$25 \text{ g dextrose} \times \frac{100 \text{ g solution}}{10 \text{ g dextrose}} = 250 \text{ g solution}$$

24. To say that the solution is 5.5% by mass $Na_2CO_3$ means that 5.5 g of $Na_2CO_3$ are contained in every 100 g of the solution.

$$500. \text{ g solution} \times \frac{5.5 \text{ g } Na_2CO_3}{100 \text{ g solution}} = 28 \text{ g } Na_2CO_3$$

26. $$100. \text{ g solution} \times \frac{1 \text{ g } AgNO_3}{100 \text{ g solution}} = 1 \text{ g } AgNO_3$$

28. For NaCl: $$125 \text{ g solution} \times \frac{7.5 \text{ g NaCl}}{100 \text{ g solution}} = 9.4 \text{ g NaCl}$$

For KBr: $$125 \text{ g solution} \times \frac{2.5 \text{ g KBr}}{100 \text{ g solution}} = 3.1 \text{ g KBr}$$

## 15.4 Solution Composition: Molarity

30. 2.0 mol $Na^+$; 1.0 mol $SO_4^{2-}$

32. To say that a solution has a concentration of 5 *M* means that in 1 L of solution (*not* solvent) there would be 5 mol of solute: to prepare such a solution one would place 5 mol of NaCl in a 1 L flask, and then add whatever amount of water is necessary so that the *total* volume would be 1 L after mixing.

34. $$\text{Molarity} = \frac{\text{moles of solute}}{\text{liters of solution}}$$

a. 25 mL = 0.025 L

$$M = \frac{0.10 \text{ mol } CaCl_2}{0.025 \text{ L solution}} = 4.0\ M$$

b. $$M = \frac{2.5 \text{ mol KBr}}{2.5 \text{ L solution}} = 1.0\ M$$

c. 755 mL = 0.755 L

$$M = \frac{0.55 \text{ mol } NaNO_3}{0.755 \text{ L solution}} = 0.73\ M$$

d. $$M = \frac{4.5 \text{ mol } Na_2SO_4}{1.25 \text{ L solution}} = 3.6\ M$$

36. $$\text{Molarity} = \frac{\text{moles of solute}}{\text{liters of solution}}$$

a. molar mass $BaCl_2$ = 208.2 g

$$5.0 \text{ g } BaCl_2 \times \frac{1 \text{ mol}}{208.2 \text{ g}} = 0.0240 \text{ mol } BaCl_2$$

$$M = \frac{0.240 \text{ mol } BaCl_2}{2.5 \text{ L solution}} = 9.6 \times 10^{-3}\ M$$

b. molar mass KBr = 119.0 g

$$3.5 \text{ g KBr} \times \frac{1 \text{ mol}}{119.0 \text{ g}} = 0.0294 \text{ mol KBr}$$

75 mL = 0.075 L

$$M = \frac{0.0294 \text{ mol KBr}}{0.075 \text{ L solution}} = 0.39\ M$$

c. molar mass $Na_2CO_3$ = 106.0 g

$$21.5 \text{ g } Na_2CO_3 \times \frac{1 \text{ mol}}{106.0 \text{ g}} = 0.2028 \text{ mol } Na_2CO_3$$

175 mL = 0.175 L

$$M = \frac{0.2028 \text{ mol } Na_2CO_3}{0.175 \text{ L solution}} = 1.16\ M$$

d. molar mass $CaCl_2$ = 111.0 g

$$55 \text{ g } CaCl_2 \times \frac{1 \text{ mol}}{111.0 \text{ g}} = 0.495 \text{ mol } CaCl_2$$

$$M = \frac{0.495 \text{ mol } CaCl_2}{1.2 \text{ L solution}} = 0.41\ M$$

38. molar mass $C_{12}H_{22}O_{11}$ = 342.3 g

$$125 \text{ g } C_{12}H_{22}O_{11} \times \frac{1 \text{ mol}}{342.3 \text{ g}} = 0.3652 \text{ mol } C_{12}H_{22}O_{11}$$

450. mL = 0.450 L

$$M = \frac{0.3652 \text{ mol } C_{12}H_{22}O_{11}}{0.450 \text{ L solution}} = 0.812\ M$$

40. molar mass HCl = 36.46 g

$$439 \text{ g HCl} \times \frac{1 \text{ mol}}{36.46 \text{ g}} = 12.04 \text{ mol HCl}$$

$$M = \frac{12.04 \text{ mol HCl}}{1.00 \text{ L solution}} = 12.0\ M$$

42. molar mass NaCl = 58.44 g

$$1.5 \text{ g NaCl} \times \frac{1 \text{ mol}}{58.44 \text{ g}} = 0.0257 \text{ mol NaCl}$$

$$M = \frac{0.0257 \text{ mol NaCl}}{1.0 \text{ L solution}} = 0.026\ M$$

44. $\text{Molarity} = \dfrac{\text{moles of solute}}{\text{liters of solution}}$

a. $1.5 \text{ L solution} \times \dfrac{3.0 \text{ mol } H_2SO_4}{1.00 \text{ L solution}} = 4.5 \text{ mol } H_2SO_4$

b. $35 \text{ mL} = 0.035 \text{ L}$

$0.035 \text{ L solution} \times \dfrac{5.4 \text{ mol NaCl}}{1.00 \text{ L solution}} = 0.19 \text{ mol NaCl}$

c. $5.2 \text{ L solution} \times \dfrac{18 \text{ mol } H_2SO_4}{1.00 \text{ L solution}} = 94 \text{ mol } H_2SO_4$

d. $0.050 \text{ L} \times \dfrac{1.1 \times 10^{-3} \text{ mol NaF}}{1.00 \text{ L solution}} = 5.5 \times 10^{-5} \text{ mol NaF}$

46. a. $3.8 \text{ L solution} \times \dfrac{1.5 \text{ mol KCl}}{1.00 \text{ L solution}} = 5.7 \text{ mol KCl}$

molar mass KCl = 74.6 g

$5.7 \text{ mol KCl} \times \dfrac{74.6 \text{ g KCl}}{1 \text{ mol KCl}} = 420 \text{ g KCl} = 4.2 \times 10^2 \text{ g KCl}$

b. $15 \text{ mL} = 0.015 \text{ L}$

$0.015 \text{ L solution} \times \dfrac{5.4 \text{ mol NaCl}}{1.00 \text{ L solution}} = 0.0810 \text{ mol NaCl}$

molar NaCl = 58.44 g

$0.0810 \text{ mol NaCl} \times \dfrac{58.44 \text{ g NaCl}}{1 \text{ mol NaCl}} = 4.7 \text{ g NaCl}$

c. $20. \text{ L solution} \times \dfrac{12.1 \text{ mol HCl}}{1.00 \text{ L solution}} = 242 \text{ mol HCl}$

molar mass HCl = 36.46 g

$242 \text{ mol HCl} \times \dfrac{36.46 \text{ g HCl}}{1 \text{ mol HCl}} = 8823 \text{ g HCl} = 8.8 \times 10^3 \text{ g HCl}$

d. $25 \text{ mL} = 0.025 \text{ L}$

$0.025 \text{ L solution} \times \dfrac{0.100 \text{ mol } HClO_4}{1.00 \text{ L solution}} = 0.00250 \text{ mol } HClO_4$

molar mass $HClO_4$ = 100.5 g

$$0.00250 \text{ mol } HClO_4 \times \frac{100.5 \text{ g } HClO_4}{1 \text{ mol } HClO_4} = 0.25 \text{ g } HClO_4$$

48. molar mass $AgNO_3$ = 169.9 g

$$10. \text{ g } AgNO_3 \times \frac{1 \text{ mol } AgNO_3}{169.9 \text{ g } AgNO_3} = 0.0589 \text{ mol } AgNO_3$$

$$0.0589 \text{ mol } AgNO_3 \times \frac{1.00 \text{ L solution}}{0.25 \text{ mol } AgNO_3} = 0.24 \text{ L solution}$$

50. a. $$1.25 \text{ L} \times \frac{0.250 \text{ mol } Na_3PO_4}{1.00 \text{ L}} = 0.3125 \text{ mol } Na_3PO_4 =$$

$$0.3125 \text{ mol } Na_3PO_4 \times \frac{3 \text{ mol } Na^+}{1 \text{ mol } Na_3PO_4} = 0.938 \text{ mol } Na^+$$

$$0.3125 \text{ mol } Na_3PO_4 \times \frac{1 \text{ mol } PO_4^{3-}}{1 \text{ mol } Na_3PO_4} = 0.313 \text{ mol } PO_4^{3-}$$

b. 3.5 mL = 0.0035 L

$$0.0035 \text{ L} \times \frac{6.0 \text{ mol } H_2SO_4}{1.00 \text{ L}} = 0.021 \text{ mol } H_2SO_4$$

$$0.021 \text{ mol } H_2SO_4 \times \frac{2 \text{ mol } H^+}{1 \text{ mol } H_2SO_4} = 0.042 \text{ mol } H^+$$

$$0.021 \text{ mol } H_2SO_4 \times \frac{1 \text{ mol } SO_4^{2-}}{1 \text{ mol } H_2SO_4} = 0.021 \text{ mol } SO_4^{2-}$$

c. 25 mL = 0.025 L

$$0.025 \text{ L} \times \frac{0.15 \text{ mol } AlCl_3}{1.00 \text{ L}} = 0.00375 \text{ mol } AlCl_3$$

$$0.00375 \text{ mol } AlCl_3 \times \frac{1 \text{ mol } Al^{3+}}{1 \text{ mol } AlCl_3} = 0.0038 \text{ mol } Al^{3+}$$

$$0.00375 \text{ mol } AlCl_3 \times \frac{3 \text{ mol } Cl^-}{1 \text{ mol } AlCl_3} = 0.011 \text{ mol } Cl^-$$

d. $$1.50 \text{ L} \times \frac{1.25 \text{ mol } BaCl_2}{1.00 \text{ L}} = 1.875 \text{ mol } BaCl_2$$

$$1.875 \text{ mol } BaCl_2 \times \frac{1 \text{ mol } Ba^{2+}}{1 \text{ mol } BaCl_2} = 1.88 \text{ mol } Ba^{2+}$$

$$1.875 \text{ mol } BaCl_2 \times \frac{2 \text{ mol } Cl^-}{1 \text{ mol } BaCl_2} = 3.75 \text{ mol } Cl^-$$

52. $500. \text{ mL} = 0.500 \text{ L}$

$$0.500 \text{ L} \times \frac{0.0200 \text{ mol } CaCO_3}{1.00 \text{ L}} = 0.0100 \text{ mol } CaCO_3 \text{ needed}$$

molar mass $CaCO_3 = 100.1$ g

$$0.0100 \text{ mol } CaCO_3 \times \frac{100.1 \text{ g } CaCO_3}{1 \text{ mol } CaCO_3} = 1.00 \text{ g } CaCO_3$$

## 15.5 Dilution

54. half

56. $M_1 \times V_1 = M_2 \times V_2$

a. $M_1 = 0.200\ M$ $\quad M_2 = ?$

$V_1 = 125$ mL $\quad V_2 = 125 + 150. = 275$ mL

$$M_2 = \frac{(0.200\ M)(125 \text{ mL})}{(275 \text{ mL})} = 0.0909\ M$$

b. $M_1 = 0.250\ M$ $\quad M_2 = ?$

$V_1 = 155$ mL $\quad V_2 = 155 + 150. = 305$ mL

$$M_2 = \frac{(0.250\ M)(155 \text{ mL})}{(305 \text{ mL})} = = 0.127\ M$$

c. $M_1 = 0.250\ M$ $\quad M_2 = ?$

$V_1 = 0.500 \text{ L} = 500. \text{ mL}$ $\quad V_2 = 500. + 150. = 650.$ mL

$$M_2 = \frac{(0.250 \text{ M})(500. \text{ mL})}{(650 \text{ mL})} = 0.192\ M$$

d. $M_1 = 18.0\ M$ $\quad M_2 = ?$

$V_1 = 15$ mL $\quad V_2 = 15 + 150. = 165$ mL

$$M_2 = \frac{(18.0\ M)(15 \text{ mL})}{(165 \text{ mL})} = 1.6\ M$$

58. $M_1 \times V_1 = M_2 \times V_2$

$M_1 = 12.1\ M$ $M_2 = 0.100\ M$

$V_1 = ?$ $V_2 = 100.\ \text{mL}$

$$V_1 = \frac{(0.100\ M)(100.\ \text{mL})}{(12.1\ M)} = 0.826\ \text{mL}$$

60. $M1 \times V_1 = M_2 \times V_2$

$M1 = 5.4\ M$ $M_2 = ?$

$V_1 = 50.\ \text{mL}$ $V_2 = 300.\ \text{mL}$

$$M_2 = \frac{(5.4\ M)(50.\ \text{mL})}{(300.\ \text{mL})} = 0.90\ \text{M}$$

62. $M_1 \times V_1 = M_2 \times V_2$

$M_1 = 6.0\ M$ $M_2 = ?$

$V_1 = 3.0\ \text{L}$ $V_2 = 10.0 + 3.0 = 13.0\ \text{L}$

$$M_2 = \frac{(6.0\ M)(3.0\ \text{L})}{(13.0\ \text{L})} = 1.4\ M$$

## 15.6 Stoichiometry of Solution Reactions

64. 25.0 mL = 0.0250 L

$$0.0250\ \text{L}\ NiCl_2\ \text{solution} \times \frac{0.20\ \text{mol}\ NiCl_2}{1.00\ \text{L}\ NiCl_2\ \text{solution}} = 0.00500\ \text{mol}\ NiCl_2$$

$$0.00500\ \text{mol}\ NiCl_2 \times \frac{1\ \text{mol}\ Na_2S}{1\ \text{mol}\ NiCl_2} = 0.00500\ \text{mol}\ Na_2S$$

$$0.00500\ \text{mol}\ Na_2S \times \frac{1.00\ \text{L}\ Na_2S\ \text{solution}}{0.10\ \text{mol}\ Na_2S} = 0.050\ \text{L} = 50.\ \text{mL}\ Na_2S\ \text{solution}$$

66. 15.3 mL = 0.0153 L

$$0.0153\ \text{L} \times \frac{0.139\ \text{mol}\ H_2SO_4}{1.00\ \text{L}} = 2.127 \times 10^{-3}\ \text{mol}\ H_2SO_4$$

$$2.127 \times 10^{-3}\ \text{mol}\ H_2SO_4 \times \frac{1\ \text{mol}\ Ba(NO_3)_2}{1\ \text{mol}\ H_2SO_4} = 2.127 \times 10^{-3}\ \text{mol}\ Ba(NO_3)_2$$

molar mass $Ba(NO_3)_2$ = 261.3 g

$$2.127 \times 10^{-3} \text{ mol } Ba(NO_3)_2 \times \frac{261.\ 3 \text{ g } Ba(NO_3)_2}{1 \text{ mol } Ba(NO_3)_2} = 0.556 \text{ g } Ba(NO_3)_2$$

68. 10.0 mL = 0.0100 L

$$0.0100 \text{ L} \times \frac{0.250 \text{ mol } AlCl_3}{1.00 \text{ L}} = 2.50 \times 10^{-3} \text{ mol } AlCl_3$$

$AlCl_3(aq) + 3NaOH(s) \rightarrow Al(OH)_3(s) + 3NaCl(aq)$

$$2.50 \times 10^{-3} \text{ mol } AlCl_3 \times \frac{3 \text{ mol NaOH}}{1 \text{ mol } AlCl_3} = 7.50 \times 10^{-3} \text{ mol NaOH}$$

molar mass NaOH = 40.0 g

$$7.50 \times 10^{-3} \text{ mol NaOH} \times \frac{40.0 \text{ g NaOH}}{1 \text{ mol}} = 0.300 \text{ g NaOH}$$

## 15.7 Neutralization Reactions

70. $HNO_3(aq) + NaOH(aq) \rightarrow NaNO_3(aq) + H_2O(l)$

35.0 mL = 0.0350 L

$$0.0350 \text{ L} \times \frac{0.150 \text{ mol NaOH}}{1.00 \text{ L}} = 5.25 \times 10^{-3} \text{ mol NaOH}$$

$$5.25 \times 10^{-3} \text{ mol NaOH} \times \frac{1 \text{ mol } HNO_3}{1 \text{ mol NaOH}} = 5.25 \times 10^{-3} \text{ mol } HNO_3$$

$$5.25 \times 10^{-3} \text{ mol } HNO_3 \times \frac{1.00 \text{ L}}{0.150 \text{ mol } HNO_3} = 0.0350 \text{ L} = 35.0 \text{ mL } HNO_3$$

72. 7.2 mL = 0.0072 L

$$0.0072 \text{ L} \times \frac{2.5 \times 10^{-3} \text{ mol NaOH}}{1.00 \text{ L}} = 1.8 \times 10^{-5} \text{ mol NaOH}$$

$H^+(aq) + OH^-(aq) \rightarrow H_2O(l)$

$$1.8 \times 10^{-5} \text{ mol } OH^- \times \frac{1 \text{ mol } H^+}{1 \text{ mol } OH^-} = 1.8 \times 10^{-5} \text{ mol } H^+$$

100 mL = 0.100 L

$$M = \frac{1.8 \times 10^{-5} \text{ mol } H^+}{0.100 \text{ L}} = 1.8 \times 10^{-4}\ M\ H^+(aq)$$

74\. a. $HCl(aq) + NaOH(aq) \rightarrow NaCl(aq) + H_2O(l)$

25.0 mL = 0.0250 L

$$0.0250 \text{ L} \times \frac{0.103 \text{ mol NaOH}}{1.00 \text{ L}} = 0.02575 \text{ mol NaOH}$$

$$0.02575 \text{ mol NaOH} \times \frac{1 \text{ mol HCl}}{1 \text{ mol NaOH}} = 0.02575 \text{ mol HCl}$$

$$0.02575 \text{ mol HCl} \times \frac{1.00 \text{ L}}{0.250 \text{ mol HCl}} = 0.0103 \text{ L HCl} = 10.3 \text{ mL HCl}$$

b. $2HCl(aq) + Ca(OH)_2(aq) \rightarrow CaCl_2(aq) + 2H_2O(l)$

50.0 mL = 0.0500 L

$$0.0500 \text{ L} \times \frac{0.00501 \text{ mol Ca(OH)}_2}{1.00 \text{ L}} = 2.505 \times 10^{-4} \text{ mol Ca(OH)}_2$$

$$2.505 \times 10^{-4} \text{ mol Ca(OH)}_2 \times \frac{2 \text{ mol HCl}}{1 \text{ mol Ca(OH)}_2} = 5.010 \times 10^{-4} \text{ mol HCl}$$

$$5.010 \times 10^{-4} \text{ mol HCl} \times \frac{1.00 \text{ L}}{0.250 \text{ mol HCl}} = 0.00200 \text{ L} = 2.00 \text{ mL}$$

c. $HCl(aq) + NH_3(aq) \rightarrow NH_4Cl(aq)$

20.0 mL = 0.0200 L

$$0.0200 \text{ L} \times \frac{0.226 \text{ mol NH}_3}{1.00 \text{ L}} = 0.00452 \text{ mol NH}_3$$

$$0.00452 \text{ mol NH}_3 \times \frac{1 \text{ mol HCl}}{1 \text{ mol NH}_3} = 0.00452 \text{ mol HCl}$$

$$0.00452 \text{ mol HCl} \times \frac{1.00 \text{ L}}{0.250 \text{ mol HCl}} = 0.01808 \text{ L} = 18.1 \text{ mL}$$

d. $HCl(aq) + KOH(aq) \rightarrow KCl(aq) + H_2O(l)$

15.0 mL = 0.0150 L

$$0.0150 \text{ L} \times \frac{0.0991 \text{ mol KOH}}{1.00 \text{ L}} = 1.487 \times 10^{-3} \text{ mol KOH}$$

$$1.487 \times 10^{-3} \text{ mol KOH} \times \frac{1 \text{ mol HCl}}{1 \text{ mol KOH}} = 1.487 \times 10^{-3} \text{ mol HCl}$$

$$1.487 \times 10^{-3} \text{ mol HCl} \times \frac{1.00 \text{ L}}{0.250 \text{ mol HCl}} = 0.00595 \text{ L} = 5.95 \text{ mL}$$

## 15.8 Solution Composition: Normality

76. 1 normal

78. 1.53 equivalents $OH^-$ ion. By definition, one equivalent of $OH^-$ ion exactly neutralizes one equivalent of $H^+$ ion.

80. $$N = \frac{\text{number of equivalents of solute}}{\text{number of liters of solution}}$$

a. equivalent weight HCl = molar mass HCl = 36.46 g

$$15.0\ \text{g HCl} \times \frac{1\ \text{equiv HCl}}{36.46\ \text{g HCl}} = 0.411\ \text{equiv HCl}$$

500. mL = 0.500 L

$$N = \frac{0.411\ \text{equiv}}{0.500\ \text{L}} = 0.822\ N$$

b. $$\text{equivalent weight } H_2SO_4 = \frac{\text{molar mass}}{2} = \frac{98.0\ \text{g}}{2} = 49.0\ \text{g}$$

$$49.0\ \text{g } H_2SO_4 \times \frac{1\ \text{equiv } H_2SO_4}{49.0\ \text{g } H_2SO_4} = 1.00\ \text{equiv } H_2SO_4$$

250. mL = 0.250 L

$$N = \frac{1.00\ \text{equiv}}{0.250\ \text{L}} = 4.00\ N$$

c. $$\text{equivalent weight } H_3PO_4 = \frac{\text{molar mass}}{3} = \frac{98.0\ \text{g}}{3} = 32.67\ \text{g}$$

$$10.0\ \text{g } H_3PO_4 \times \frac{1\ \text{equiv } H_3PO_4}{32.67\ \text{g } H_3PO_4} = 0.3061\ \text{equiv } H_3PO_4$$

100. mL = 0.100 L

$$N = \frac{0.3061\ \text{equiv}}{0.100\ \text{L}} = 3.06\ N$$

82. a. $$0.50\ M\ HC_2H_3O_2 \times \frac{1\ \text{equiv } HC_2H_3O_2}{1\ \text{mol } HC_2H_3O_2} = 0.50\ N\ HC_2H_3O_2$$

b. $$0.00250\ M\ H_2SO_4 \times \frac{2\ \text{equiv } H_2SO_4}{1\ \text{mol } H_2SO_4} = 0.00500\ N\ H_2SO_4$$

c. $$0.10\ M\ \text{KOH} \times \frac{1\ \text{equiv KOH}}{1\ \text{mol KOH}} = 0.10\ N\ \text{KOH}$$

84. molar mass $NaH_2PO_4$ = 120.0 g

$$5.0\ \text{g NaH}_2\text{PO}_4 \times \frac{1\ \text{mol NaH}_2\text{PO}_4}{120.0\ \text{g NaH}_2\text{PO}_4} = 0.04166\ \text{mol NaH}_2\text{PO}_4$$

500. mL = 0.500 L

$$M = \frac{0.04166\ \text{mol}}{0.500\ \text{L}} = 0.08333\ M\ \text{NaH}_2\text{PO}_4 = 0.083\ M\ \text{NaH}_2\text{PO}_4$$

$$0.08333\ M\ \text{NaH}_2\text{PO}_4 \times \frac{2\ \text{equiv NaH}_2\text{PO}_4}{1\ \text{mol NaH}_2\text{PO}_4} = 0.1667\ N\ \text{NaH}_2\text{PO}_4 = 0.17\ N\ \text{NaH}_2\text{PO}_4$$

86. $2NaOH(aq) + H_2SO_4(aq) \rightarrow Na_2SO_4(aq) + 2H_2O(l)$

15.0 mL = 0.0150 L

$$0.0150\ \text{L} \times \frac{0.35\ \text{mol H}_2\text{SO}_4}{1.00\ \text{L}} = 0.00525\ \text{mol H}_2\text{SO}_4$$

$$0.00525\ \text{mol H}_2\text{SO}_4 \times \frac{2\ \text{mol NaOH}}{1\ \text{mol H}_2\text{SO}_4} = 0.0105\ \text{mol NaOH}$$

$$0.0105\ \text{mol NaOH} \times \frac{1.00\ \text{L}}{0.50\ \text{mol NaOH}} = 0.0210\ \text{L} = 21\ \text{mL}$$

88. $N_{\text{acid}} \times V_{\text{acid}} = N_{\text{base}} \times V_{\text{base}}$

$N_{\text{acid}} \times (10.0\ \text{mL}) = (3.5 \times 10^{-2}\ N)(27.5\ \text{mL})$

$N_{\text{acid}} = 9.6 \times 10^{-2}\ N\ \text{HNO}_3$

## Additional Problems

90. Molarity is defined as the number of moles of solute contained in 1 liter of *total* solution volume (solute plus solvent after mixing). In the first case, where 50. g of NaCl is dissolved in 1.0 L of water, the total volume after mixing is *not* known and the molarity cannot be calculated. In the second example, the final volume after mixing is known and the molarity can be calculated simply.

92. $$75\ \text{g solution} \times \frac{25\ \text{g NaCl}}{100\ \text{g solution}} = 18.75\ \text{g NaCl}$$

$$18.75\ \text{g NaCl} \times \frac{100\ \text{g solution}}{5.0\ \text{g NaCl}} = 375\ \text{g} = 3.8 \times 10^2\ \text{g}$$

94. $Ba(NO_3)_2(aq) + H_2SO_4(aq) \rightarrow BaSO_4(s) + 2HNO_3(aq)$

10. mL = 0.010 L

$$0.010 \text{ L } Ba(NO_3)_2 \times \frac{0.50 \text{ mol } Ba(NO_3)_2}{1.00 \text{ L}} = 0.0050 \text{ mol } Ba(NO_3)_2$$

$$0.010 \text{ L } H_2SO_4 \times \frac{0.20 \text{ mol } H_2SO_4}{1.00 \text{ L}} = 0.0020 \text{ mol } H_2SO_4$$

Since the coefficients of $Ba(NO_3)_2$ and $H_2SO_4$ in the balanced chemical equation for the reaction are both *one*, then the 0.0020 mol $H_2SO_4$ is the limiting reactant for the precipitation reaction. Only 0.0020 mol of $Ba^{2+}$ ion will precipitate as $BaSO_4(s)$, leaving (0.0050 − 0.0020) = 0.0030 mol of $Ba^{2+}$ ion remaining in solution. This quantity of $Ba^{2+}$ ion left in solution is now dispersed through the *combined* volume of solution (20 mL = 0.020 L).

molar mass $BaSO_4$ = 233.4 g

$$0.0020 \text{ mol } BaSO_4 \times \frac{233.4 \text{ g } BaSO_4}{1 \text{ mol } BaSO_4} = 0.47 \text{ g } BaSO_4 \text{ precipitate}$$

$$M_{\text{barium remaining}} = \frac{0.0030 \text{ mol } Ba^{2+}}{0.020 \text{ L}} = 0.15\ M\ Ba^{2+} \text{ in solution}$$

96. molar mass $H_2O$ = 18.0 g

1.0 L water = $1.0 \times 10^3$ mL water ≈ $1.0 \times 10^3$ g water

$$1.0 \times 10^3 \text{ g } H_2O \times \frac{1 \text{ mol } H_2O}{18.0 \text{ g } H_2O} = 56 \text{ mol } H_2O$$

98. 500 mL HCl solution = 0.500 L HCl solution

$$0.500 \text{ L solution} \times \frac{0.100 \text{ mol HCl}}{1.00 \text{ L HCl solution}} = 0.0500 \text{ mol HCl}$$

$$0.0500 \text{ mol HCl} \times \frac{22.4 \text{ L HCl gas at STP}}{1 \text{ mol HCl}} = 1.12 \text{ L HCl gas at STP}$$

100. $$10.0 \text{ g HCl} \times \frac{100.\text{ g solution}}{33.1 \text{ g HCl}} = 30.21 \text{ g solution}$$

$$30.21 \text{ g solution} \times \frac{1.00 \text{ mL solution}}{1.147 \text{ g solution}} = 26.3 \text{ mL solution}$$

102. molar mass NaCl = 58.44 g

$$10.0 \text{ g NaCl} \times \frac{1 \text{ mol NaCl}}{58.44 \text{ g NaCl}} = 0.1711 \text{ mol NaCl}$$

100. mL = 0.100 L

$$M = \frac{0.1711 \text{ mol NaCl}}{0.100 \text{ L}} = 1.71\ M$$

104. molar mass $CH_3CH_2OH$ = 46.1 g

$$50.\text{ mL} \times \frac{0.80 \text{ g } CH_3CH_2OH}{1.0 \text{ mL}} = 40.\text{ g } CH_3CH_2OH$$

$$40.\text{ g } CH_3CH_2OH \times \frac{1 \text{ mol } CH_3CH_2OH}{46.1 \text{ g } CH_3CH_2OH} = 0.870 \text{ mol } CH_3CH_2OH$$

95 mL = 0.095 L

$$M = \frac{0.870 \text{ mol } CH_3CH_2OH}{0.095 \text{ L solution}} = 9.2\ M$$

106. a. $$0.10\ M\ FeCl_3 \times \frac{3 \text{ mol } Cl^-}{1 \text{ mol } FeCl_3} = 0.30\ M\ Cl^-$$

b. $$1.0 \times 10^{-3}\ M\ AuCl_4^- \times \frac{4 \text{ mol } Cl^-}{1 \text{ mol } AuCl_4^-} = 4.0 \times 10^{-3}\ M\ Cl^-$$

c. $$2.0\ M\ CoCl_2 \times \frac{2 \text{ mol } Cl^-}{1 \text{ mol } CoCl_2} = 4.0\ M\ Cl^-$$

108. molar mass $Ca(OH)_2$ = 74.1 g

$$1.04 \text{ g } Ca(OH)_2 \times \frac{1 \text{ mol } Ca(OH)_2}{74.1 \text{ g } Ca(OH)_2} = 0.01404 \text{ mol } Ca(OH)_2$$

500. mL = 0.500 L

$$M = \frac{0.01404 \text{ mol } Ca(OH)_2}{0.500 \text{ L}} = 0.02807\ M = 0.0281\ M$$

$$N = 0.2807 \frac{\text{mol}}{\text{L}} \times \frac{2 \text{ equiv}}{1 \text{ mol}} = 0.0561 \frac{\text{equiv}}{\text{L}} = 0.0561\ N$$

# Chapter Sixteen Equilibrium

## 16.1 How Chemical Reactions Occur

2. A higher concentration means there are more molecules present, which results in a greater frequency of collision between molecules.

## 16.2 Conditions That Affect Reaction Rates

4. At higher temperatures, the average kinetic energy of the reactant molecules is larger. At higher temperatures, the probability that a collision between molecules will be energetic enough for reaction to take place is larger. On a molecular basis, a higher temperature means a given molecule will be moving faster.

6. $Cl + O_3 \rightarrow ClO + O_2$

   $O + ClO \rightarrow Cl + O_2$

   Most chlorine atoms in the upper atmosphere come from man-made sources (such as the decomposition of freons such as $CF_2Cl_2$). The ozone layer of the atmosphere is important because ozone is able to absorb harmful ultraviolet radiation from the sun. Although it may already be too late, it is likely that the ozone layer of the atmosphere will be replenished if the use of chlorinated aerosol compounds is stopped.

## 16.3 The Equilibrium Condition

8. When a liquid is confined in an otherwise empty closed container, the liquid begins to evaporate, producing molecules of vapor in the empty space of the container. As the amount of vapor increases, molecules in the vapor phase begin to condense and reenter the liquid state. Eventually the opposite processes of evaporation and condensation will be going on at the same speed: beyond this point, for every molecule that leaves the liquid state and evaporates, there is a molecule of vapor which leaves the vapor state and condenses. We know the state of equilibrium has been reached when there is no further change in the pressure of the vapor.

10. Chemical equilibrium occurs when two opposing chemical reactions reach the same speed in a closed system. When a state of chemical equilibrium has been reached, the concentrations of reactants and products present in the system remain constant with time. A chemical reaction that reaches a state of equilibrium is indicated by using a *double arrow* ($\rightleftharpoons$). The points of the double arrow point in opposite directions, indicating that two opposite processes are going on.

## 16.4 Chemical Equilibrium: A Dynamic Condition

12. Although we recognize a state of chemical equilibrium by the fact that the concentrations of reactants and products no longer change with time, the lack of change results from the fact that two opposing processes are going on at the same time (not because the reaction has "stopped"). Further reaction in the forward direction is cancelled out by an equal

extent of reaction in the reverse direction. The reaction is still proceeding, but the opposite reaction is also proceeding at the same rate.

## 16.5 The Equilibrium Constant: An Introduction

14. The equilibrium constant is a *ratio* of the concentration of products to the concentration of reactants, all at equilibrium. Depending on how much reactant a particular experiment was begun with, there may be different amounts of reactants and products present at equilibrium, but the *ratio* will always be the same for a given reaction at a given temperature. For example, the ratios 4/2 and 6/3 are different absolutely in terms of the numbers involved, but each of these ratios has the value 2.

16. a.

$$K = \frac{[HBr(g)]^2}{[H_2(g)][Br_2(g)]}$$

b.

$$K = \frac{[H_2S(g)]^2}{[H_2(g)]^2[S_2(g)]}$$

c.

$$K = \frac{[HCN(g)]^2}{[H_2(g)][C_2N_2(g)]}$$

18 a.

$$K = \frac{[O_2(g)]^3}{[O_3(g)]^2}$$

b.

$$K = \frac{[CO_2(g)][H_2O(g)]^2}{[CH_4(g)][O_2(g)]^2}$$

c.

$$K = \frac{[C_2H_4Cl_2(g)]}{[C_2H_4(g)][Cl_2(g)]}$$

20.

$$K = \frac{[Br(g)]^2}{[Br_2(g)]} = \frac{(0.034\ M)^2}{(0.97\ M)} = 1.2 \times 10^{-3}$$

22.

$$K = \frac{[PCl_3(g)][Cl_2(g)]}{[PCl_5(g)]} = \frac{(0.325\ M)(3.9 \times 10^{-3}\ M)}{(1.1 \times 10^{-2}\ M)} = 0.12$$

## 16.6 Heterogeneous Equilibria

24. Equilibrium constants represent ratios of the *concentrations* of products and reactants present at the point of equilibrium. The *concentration* of a pure solid or of a pure liquid is constant and is determined by the density of the solid or liquid.

26. a. $$K = \frac{1}{[O_2(g)]^3}$$

b. $$K = \frac{1}{[NH_3(g)][HCl(g)]}$$

c. $$K = \frac{1}{[O_2(g)]}$$

28. a. $$K = \frac{1}{[O_2(g)]^5}$$

b. $$K = \frac{[H_2O(g)]}{[CO_2(g)]}$$

c. $K = [N_2O(g)][H_2O(g)]^2$

## 16.7 Le Châtelier's Principle

30. When an additional amount of one of the reactants is added to an equilibrium system, the system shifts to the right and adjusts so as to use up some of the added reactant. This results in a net *increase* in the amount of product, compared to the equilibrium system before the additional reactant was added. The numerical *value* of the equilibrium constant does *not* change when a reactant is added: the concentrations of all reactants and products adjust until the correct value of $K$ is once again achieved.

32. An *exo*thermic reaction is one which liberates heat energy. Increasing the temperature (adding heat) for such a reaction is fighting against the reaction's own tendency to liberate heat. The net effect of raising the temperature will be a shift to the left to decrease the amount of product. If it is desired to increase the amount of products in an exothermic reaction, heat must be *removed* from the system. Changing the temperature *does* change the numerical value of the equilibrium constant for a reaction.

34. $2NO(g) + O_2(g) \rightleftharpoons 2NO_2(g)$

a. shifts to right

b. shifts to right

c. no effect (He is not involved in the reaction)

36. $UO_2(s) + 4HF(g) \rightleftharpoons UF_4(g) + 2H_2O(g)$

a. no effect ($UO_2$ is a solid)

b. no effect (Xe is not involved in the reaction)

c. shifts to left (if HF attacks glass, it is removed from system, causing reaction to replace the lost HF)

d. shifts to right

e. shifts to left (4 mol gas versus 3 mol gas)

38. The reaction is *exo*thermic as written. An increase in temperature (addition of heat) will shift the reaction to the left (toward reactants).

40. For an *endo*thermic reaction, an increase in temperature will shift the position of equilibrium to the right (toward products).

42. The reaction is *exo*thermic and should be performed at as low a temperature as possible (consistent with the molecule's still having sufficient energy to react).

## 16.8 Applications Involving the Equilibrium Constant

44. A small equilibrium constant means that the concentration of products is small, compared to the concentration of reactants. The position of equilibrium lies far to the left. Reactions with very small equilibrium constants are generally not very useful as a source of the products, unless Le Châtelier's principle can be applied to shift the position of equilibrium to the point where a sufficient amount of product can be isolated.

46. $$K = \frac{[CO_2(g)][H_2(g)]}{[CO(g)][H_2O(g)]} = \frac{(1.3\ M)(1.4\ M)}{(0.71\ M)(0.66\ M)} = 3.9$$

48. $$K = \frac{[NH_3(g)]^2}{[N_2(g)][H_2(g)]^3}$$

$$1.3 \times 10^{-2} = \frac{[NH_3(g)]}{(0.10\ M)(0.10\ M)^3}$$

$$[NH_3(g)]^2 = 1.3 \times 10^{-2} \times (0.10) \times (0.10)^3 = 1.3 \times 10^{-6}$$

$$[NH_3(g)] = \sqrt{(1.3 \times 10^{-6})} = 1.1 \times 10^{-3}\ M$$

50. $$K = \frac{[NO(g)]^2[Cl_2(g)]}{[NOCl(g)]^2}$$

$$9.2 \times 10^{-6} = \frac{(1.5 \times 10^{-3})^2[Cl_2(g)]}{(0.44\ M)^2}$$

$$[Cl_2(g)] = \frac{(9.2 \times 10^{-6})(0.44)^2}{(1.5 \times 10^{-3})^2} = \frac{1.781 \times 10^{-6}}{2.25 \times 10^{-6}} = 0.79\ M$$

52. $$K = \frac{[N_2O_4(g)]}{[NO_2(g)]^2}$$

$$1.2 \times 10^4 = \frac{(0.45\ M)}{[NO_2(g)]^2}$$

$$[NO_2(g)]^2 = \frac{(0.45)}{1.2 \times 10^4} = 3.75 \times 10^{-5}$$

$$[NO_2(g)] = \sqrt{(3.75 \times 10^{-5})} = 6.1 \times 10^{-3}\ M$$

## 16.9 Solubility Equilibria

54. solubility product, $K_{sp}$

56. Stirring or grinding the solute increases the speed with which the solute dissolves, but the ultimate *amount* of solute that dissolves is fixed by the equilibrium constant for the dissolving process, $K_{sp}$.

58. a. $MgCO_3(s) \rightleftharpoons Mg^{2+}(aq) + CO_3^{2-}(aq)$

$K_{sp} = [Mg^{2+}(aq)][CO_3^{2-}(aq)]$

b. $Al_2(SO_4)_3(s) \rightleftharpoons 2Al^{3+}(aq) + 3SO_4^{2-}(aq)$

$K_{sp} = [Al^{3+}(aq)]^2[SO_4^{2-}(aq)]^3$

c. $Ca(OH)_2(s) \rightleftharpoons Ca^{2+}(aq) + 2OH^-(aq)$

$K_{sp} = [Ca^{2+}(aq)][OH^-(aq)]^2$

d. $Ag_2SO_4(s) \rightleftharpoons 2Ag^+(aq) + SO_4^{2-}(aq)$

$K_{sp} = [Ag^+(aq)]^2[SO_4^{2-}(aq)]$

60. $NiS(s) \rightleftharpoons Ni^{2+}(aq) + S^{2-}(aq)$

$K_{sp} = [Ni^{2+}(aq)][S^{2-}(aq)] = (4.0 \times 10^{-5}\ M)(4.0 \times 10^{-5}\ M) = 1.6 \times 10^{-9}$

62. molar mass AgCl = 143.4 g

$$9.0 \times 10^{-4} \text{ g AgCl/L} \times \frac{1 \text{ mol AgCl}}{143.4 \text{ g AgCl}} = 6.28 \times 10^{-6} \text{ mol AgCl/L}$$

$AgCl(s) \rightleftharpoons Ag^{+}(aq) + Cl^{-}(aq)$

$K_{sp} = [Ag^{+}(aq)][Cl^{-}(aq)] = (6.28 \times 10^{-6}\ M)(6.28 \times 10^{-6}\ M) = 3.9 \times 10^{-11}$

64. $HgS(s) \rightleftharpoons Hg^{2+}(aq) + S^{2-}(aq)$

$K_{sp} = [Hg^{2+}(aq)][S^{2-}(aq)] = 1.6 \times 10^{-54}$

Let x represent the number of moles of HgS that dissolve per liter; then

$[Hg^{2+}(aq)] = x$ and $[S^{2-}(aq)] = x$

$K_{sp} = [x][x] = x^2 = 1.6 \times 10^{-54}$

$x = \sqrt{(1.6 \times 10^{-54})} = 1.3 \times 10^{-27}\ M$

molar mass HgS = 232.6 g

$$1.3 \times 10^{-27}\ M \times \frac{232.6 \text{ g}}{1 \text{ mol}} = 2.9 \times 10^{-25} \text{ g/L}$$

66. molar mass $Ni(OH)_2$ = 92.71 g

$$\frac{0.14 \text{ g } Ni(OH)_2}{1.00 \text{ L}} \times \frac{1 \text{ mol } Ni(OH)_2}{92.71 \text{ g } Ni(OH)_2} = 1.510 \times 10^{-3}\ M$$

$Ni(OH)_2(s) \rightleftharpoons Ni^{2+}(aq) + 2OH^{-}(aq)$

$K_{sp} = [Ni^{2+}(aq)][OH^{-}(aq)]^2$

If $1.510 \times 10^{-3}\ M$ of $Ni(OH)_2$ dissolves, then

$[Ni^{2+}(aq)] = 1.510 \times 10^{-3}\ M$ and

$[OH^{-}(aq)] = 2 \times (1.510 \times 10^{-3}\ M) = 3.020 \times 10^{-3}\ M$

$K_{sp} = (1.510 \times 10^{-3}\ M)(3.020 \times 10^{-3}\ M)^2 = 1.4 \times 10^{-8}$

68. $PbCl_2(s) \rightleftharpoons Pb^{2+}(aq) + 2Cl^{-}(aq)$

$K_{sp} = [Pb^{2+}(aq)][Cl^{-}(aq)]^2$

If $PbCl_2$ dissolves to the extent of $3.6 \times 10^{-2}\ M$, then

$[Pb^{2+}(aq)] = 3.6 \times 10^{-2}\ M$ and

$[Cl^{-}(aq)] = 2 \times (3.6 \times 10^{-2}) = 7.2 \times 10^{-2}\ M$

$K_{sp} = (3.6 \times 10^{-2}\ M)(7.2 \times 10^{-2}\ M)^2 = 1.9 \times 10^{-4}$

70. $Fe(OH)_3(s) \rightleftharpoons Fe^{3+}(aq) + 3OH^-(aq)$

$K_{sp} = [Fe^{3+}(aq)][OH^-(aq)]^3 = 4 \times 10^{-38}$

let x represent the number of moles of $Fe(OH)_3$ that dissolve per liter; then

$[Fe^{3+}(aq)] = x$ and $[OH^-(aq)] = 3x$

$K_{sp} = [x][3x]^3 = 27x^4 = 4 \times 10^{-38}$

$x^4 = 1.48 \times 10^{-39}$

$x = 2 \times 10^{-10}$ *M*

molar mass $Fe(OH)_3$ = 106.9 g

$$\frac{2 \times 10^{-10} \text{ mol}}{1.00 \text{ L}} \times \frac{106.9 \text{ g}}{1 \text{ mol}} = 2 \times 10^{-8} \text{ g/L}$$

## Additional Problems

72. temperature

74. catalyst

76. constant

78. dynamic

80. heterogeneous

82. position

84. exothermic

86. An equilibrium reaction may come to many *positions* of equilibrium, but at each possible position of equilibrium, the numerical value of the equilibrium constant is fulfilled. If different amounts of reactant are taken in different experiments, the *absolute amounts* of reactant and product present at the point of equilibrium reached will differ from one experiment to another, but the *ratio* that defines the equilibrium constant will be the same.

88. $PCl_5(g) \rightleftharpoons PCl_3(g) + Cl_2(g)$

$$K = \frac{[PCl_3(g)][Cl_2(g)]}{[PCl_5(g)]} = 4.5 \times 10^{-3}$$

The concentration of $PCl_5$ is to be twice the concentration of $PCl_3$:

$[PCl_5(g)] = 2 \times [PCl_3(g)]$

$$K = \frac{[PCl_3(g)][Cl_2(g)]}{2 \times [PCl_3(g)]} = 4.5 \times 10^{-3}$$

$$K = \frac{[Cl_2(g)]}{2} = 4.5 \times 10^{-3}$$

$[Cl_2(g)] = 9.0 \times 10^{-3}\ M$

90. $Ag_2CrO_4(s) \rightleftharpoons 2Ag^+(aq) + CrO_4^{2-}(aq)$

$K_{sp} = [Ag^+(aq)]^2[CrO_4^{2-}(aq)] = 9.1 \times 10^{-12}$

let x represent the number of moles of $Ag_2CrO_4$ that dissolve per liter; then

$[Ag^+(aq)] = 2x$ and $[CrO_4^{2-}(aq)] = x$

$K_{sp} = [2x]^2[x] = 9.1 \times 10^{-12}$

$4x^3 = 9.1 \times 10^{-12}$

$x^3 = 2.275 \times 10^{-12}$

$x = 1.3 \times 10^{-4}\ M$

molar mass $Ag_2CrO_4$ = 331.8 g

$$\frac{1.3 \times 10^{-4}\ \text{mol}}{1.00\ \text{L}} \times \frac{331.8\ \text{g}}{1\ \text{mol}} = 0.043\ \text{g/L}$$

92. Although a small solubility product generally implies a small solubility, comparisons of solubility based directly on $K_{sp}$ values are only valid if the salts produce the same numbers of positive and negative ions per formula when they dissolve. For example, one can compare the solubilities of AgCl(*s*) and NiS(*s*) directly using $K_{sp}$, since each salt produces one positive and one negative ion per formula when dissolved. One could not directly compare AgCl(*s*) with a salt such as $Ca_3(PO_4)_2$, however.

## Chapter Seventeen Acids and Bases

### 17.1 Acids and Bases

2. In the Arrhenius definition, an acid is a substance which produces hydrogen ions ($H^+$) when dissolved in water, whereas a base is a substance which produces hydroxide ions ($OH^-$) in aqueous solution. These definitions proved to be too restrictive since the only base permitted was hydroxide ion, and the only solvent permitted was water.

4. The substances in a conjugate acid-base pair differ by the presence or absence of a proton ($H^+$).

6. When an acid is dissolved in water, the hydronium ion ($H_3O^+$) is formed. The hydronium ion is the conjugate acid of water ($H_2O$).

8.
   a. $H_2O$ and $OH^-$ represent a conjugate acid-base pair ($H_2O$ is the acid, having one more proton than the base, $OH^-$).

   b. $H_2SO_4$ and $SO_4^{2-}$ are *not* a conjugate acid-base pair (they differ by *two* protons). The conjugate base of $H_2SO_4$ is $HSO_4^-$; the conjugate acid of $SO_4^{2-}$ is also $HSO_4^-$.

   c. $H_3PO_4$ and $H_2PO_4^-$ represent a conjugate acid-base pair ($H_3PO_4$ is the acid, having one more proton than the base $H_2PO_4^-$).

   d. $HC_2H_3O_2$ and $C_2H_3O_2^-$ represent a conjugate acid base pair ($HC_2H_3O_2$ is the acid, having one more proton than the base $C_2H_3O_2^-$).

10.
   a. $CH_3NH_2$ (base), $CH_3NH_3^+$ (acid); $H_2O$ (acid), $OH^-$ (base)
   b. $CH_3COOH$ (acid), $CH_3COO^-$ (base); $NH_3$ (base), $NH_4^+$ (acid)
   c. HF (acid), $F^-$ (base); $NH_3$ (base), $NH_4^+$ (acid)

12. The conjugate *acid* of the species indicated would have *one additional proton*:

   a. $NH_4^+$
   b. $CH_3COOH$
   c. $H_3PO_4$
   d. $H_2C_2O_4$

14. The conjugate *bases* of the species indicated would have *one less proton*:

   a. $H_2PO_4^-$
   b. $CO_3^{2-}$
   c. $F^-$
   d. $HSO_4^-$

16. When an acid ionizes in water, a proton is released to the water as an $H_3O^+$ ion:

    a. $CH_3CH_2COOH + H_2O \rightleftharpoons CH_3CH_2COO^- + H_3O^+$

    b. $NH_4^+ + H_2O \rightleftharpoons NH_3 + H_3O^+$

    c. $H_2SO_4 + H_2O \rightleftharpoons HSO_4^- + H_3O^+$

    d. $H_3PO_4 + H_2O \rightleftharpoons H_2PO_4^- + H_3O^+$

## 17.2 Acid Strength

18. To say that an acid is weak in aqueous solution means that the acid does not easily transfer protons to water (and does not fully ionize). If an acid does not lose protons easily, then the acid's anion must be a strong attractor of protons.

20. A strong acid is one which loses its protons easily and fully ionizes in water; this means that the acid's conjugate base must be poor at attracting and holding on to protons and is a relatively weak base. A weak acid is one which resists loss of its protons and does not ionize well in water; this means that the acid's conjugate base attracts and holds onto protons tightly and is a relatively strong base.

22. $H_2SO_4$ (sulfuric): $H_2SO_4 + H_2O \rightarrow HSO_4^- + H_3O^+$

    HCl (hydrochloric): $HCl + H_2O \rightarrow Cl^- + H_3O^+$

    $HNO_3$ (nitric): $HNO_3 + H_2O \rightarrow NO_3^- + H_3O^+$

    $HClO_4$ (perchloric): $HClO_4 + H_2O \rightarrow ClO_4^- + H_3O^+$

24. oxyacids: $HClO_4$, $HNO_3$, $H_2SO_4$, $CH_3COOH$, etc.

    non-oxyacids: HCl, HBr, HF, HI, HCN, etc.

26. Bases that are *weak* have relatively strong conjugate acids:

    a. $F^-$ is a relatively strong base; HF is a weak acid.

    b. $Cl^-$ is a very weak base; HCl is a strong acid.

    c. $HSO_4^-$ is a very weak base; $H_2SO_4$ is a strong acid.

    d. $NO_3^-$ is a very weak base; $HNO_3$ is a strong acid.

## 17.3 Water as an Acid and a Base

28. $H_2O \rightleftharpoons H^+(aq) + OH^-(aq)$

$K_w = [H^+(aq)][OH^-(aq)] = 1.0 \times 10^{-14}$ at 25°C

The very small numerical value of $K_w$ means pure water is not very much ionized at 25°C.

30. If $[H^+(aq)] > [OH^-(aq)]$ the solution is acidic.

If $[OH^-(aq)] > [H^+(aq)]$ the solution is basic.

If $[H^+(aq)] = [OH^-(aq)]$ the solution is neutral.

At 25°C, if $[H^+(aq)] > 1.0 \times 10^{-7}\ M$ the solution is acidic.

At 25°C, if $[OH^-(aq)] > 1.0 \times 10^{-7}\ M$ the solution is basic.

32. $K_w = [H^+(aq)][OH^-(aq)] = 1.0 \times 10^{-14}$ at 25°C

$$[H^+(aq)] = \frac{1.0 \times 10^{-14}}{[OH^-(aq)]}$$

a. $[H^+(aq)] = \dfrac{1.0 \times 10^{-14}}{4.22 \times 10^{-3}\ M} = 2.37 \times 10^{-12}\ M$; solution is basic

b. $[H^+(aq)] = \dfrac{1.0 \times 10^{-14}}{1.01 \times 10^{-13}\ M} = 9.90 \times 10^{-2}\ M$; solution is acidic

c. $[H^+(aq)] = \dfrac{1.0 \times 10^{-14}}{3.05 \times 10^{-7}\ M} = 3.28 \times 10^{-8}\ M$; solution is basic

d. $[H^+(aq)] = \dfrac{1.0 \times 10^{-14}}{6.02 \times 10^{-6}\ M} = 1.66 \times 10^{-9}\ M$; solution is basic

34. $$[OH^-(aq)] = \frac{1.0 \times 10^{-14}}{[H^+(aq)]}$$

a. $[OH^-(aq)] = \dfrac{1.0 \times 10^{-14}}{4.21 \times 10^{-7}\ M} = 2.38 \times 10^{-8}\ M$; solution is acidic

b. $[OH^-(aq)] = \dfrac{1.0 \times 10^{-14}}{0.00035\ M} = 2.9 \times 10^{-11}\ M$; solution is acidic

c. $[OH^-(aq)] = \dfrac{1.0 \times 10^{-14}}{0.00000010\ M} = 1.0 \times 10^{-7}\ M$; solution is neutral

d. $[OH^-(aq)] = \dfrac{1.0 \times 10^{-14}}{9.9 \times 10^{-6}\ M} = 1.0 \times 10^{-9}\ M$; solution is acidic

36. a. $[OH^-(aq)] = 0.0000032\ M$ is more basic

b. $[OH^-(aq)] = 1.54 \times 10^{-8}\ M$ is more basic

c. $[OH^-(aq)] = 4.02 \times 10^{-7}\ M$ is more basic

## 17.4 The pH Scale

38. household ammonia (pH 12); blood (pH 7-8); milk (pH 6-7); vinegar (pH 3); lemon juice (pH 2-3); stomach acid (pH 2).

40. The pH of a solution is defined as the *negative* of the logarithm of the hydrogen ion concentration, $pH = -\log[H^+]$. Mathematically, the *negative sign* in the definition causes the pH to *decrease* as the hydrogen ion concentration *increases*.

42. $pH = -\log[H^+(aq)]$

a. $pH = -\log[4.21 \times 10^{-7}\ M] = 6.376$; solution is acidic

b. $pH = -\log[0.00035\ M] = 3.46$; solution is acidic

c. $pH = -\log[0.00000010\ M] = 7.00$; solution is neutral

d. $pH = -\log[1.21 \times 10^{-3}\ M] = 2.917$; solution is acidic

44. $pOH = -\log[OH^-(aq)]$ $\quad pH = 14 - pOH$

a. $pOH = -\log[1.4 \times 10^{-6}\ M] = 5.85$

$pH = 14 - 5.85 = 8.15$; solution is basic

b. $pOH = -\log[9.35 \times 10^{-9}\ M] = 8.029$

$pH = 14 - 8.029 = 5.971$; solution is acidic

c. $pOH = -\log[2.21 \times 10^{-1}\ M] = 0.656$

$pH = 14 - 0.656 = 13.344$; solution is basic

d. $pOH = -\log[7.98 \times 10^{-12}\ M] = 11.098$

$pH = 14 - 11.098 = 2.902$; solution is acidic

46. pOH = 14 − pH

a. pOH = 14 − 1.02 = 12.98; solution is acidic

b. pOH = 14 − 13.4 = 0.6; solution is basic

c. pOH = 14 − 9.03 = 4.97; solution is basic

d. pOH = 14 − 7.20 = 6.80; solution is basic

48. a. $[OH^-(aq)] = \dfrac{1.0 \times 10^{-14}}{5.72 \times 10^{-4}\ M} = 1.75 \times 10^{-11}\ M$

$pOH = -\log[1.75 \times 10^{-11}\ M] = 10.757$

$pH = 14 - 10.757 = 3.243$

b. $[H^+(aq)] = \dfrac{1.0 \times 10^{-14}}{8.91 \times 10^{-5}\ M} = 1.12 \times 10^{-10}\ M$

$pH = -\log[1.12 \times 10^{-10}\ M] = 9.950$

$pOH = 14 - 9.950 = 4.050$

c. $[OH^-(aq)] = \dfrac{1.0 \times 10^{-14}}{2.87 \times 10^{-12}\ M} = 3.48 \times 10^{-3}\ M$

$pOH = -\log[3.48 \times 10^{-3}\ M] = 2.458$

$pH = 14 - 2.458 = 11.542$

d. $[H^+(aq)] = \dfrac{1.0 \times 10^{-14}}{7.22 \times 10^{-8}\ M} = 1.39 \times 10^{-7}\ M$

$pH = -\log[1.39 \times 10^{-7}\ M] = 6.859$

$pOH = 14 - 6.859 = 7.141$

50. $[H^+]$ = {inv}{log}[−pH]

a. $[H^+]$ = {inv}{log}[−8.34] = $4.6 \times 10^{-9}\ M$

b. $[H^+]$ = {inv}{log}[−5.90] = $1.3 \times 10^{-6}\ M$

c. $[H^+]$ = {inv}{log}[−2.65] = $2.2 \times 10^{-3}\ M$

d. $[H^+]$ = {inv}{log}[−12.6] = $3 \times 10^{-13}\ M$

52. pH = 14 − pOH $[H^+]$ = {inv}{log}[−pH]

a. pH = 14 − 7.95 = 6.05

$[H^+]$ = {inv}{log}[−6.05] = $8.9 \times 10^{-7}$ *M*

b. pH = 14 − 14.00 = 0.00

$[H^+]$ = {inv}{log}[0.00] = 1.0 *M*

c. pH = 14 − 5.00 = 9.00

$[H^+]$ = {inv}{log}[−9.00] = $1.0 \times 10^{-9}$ *M*

d. pH = 14 − 3.95 = 10.05

$[H^+]$ = {inv}{log}[−10.05] = $8.9 \times 10^{-11}$ *M*

54. a. pH = 14 − 0.90 = 13.10

$[H^+]$ = {inv}{log}[−13.10] = $7.9 \times 10^{-14}$ *M*

b. $[H^+]$ = {inv}{log}[−0.90] = 0.13 *M*

c. pH = 14 − 10.3 = 3.7

$[H^+]$ = {inv}{log}[−3.7] = $2 \times 10^{-4}$ *M*

d. $[H^+]$ = {inv}{log}[−5.33] = $4.7 \times 10^{-6}$ *M*

## 17.5 Calculating the pH of Strong Acid Solutions

56. The solution contains water molecules, $H_3O^+$ ions (protons), and $NO_3^-$ ions. Because $HNO_3$ is a strong acid, which is completely ionized in water, there are no $HNO_3$ molecules present.

58. a. $HClO_4$ is a strong acid and completely ionized so

$[H^+]$ = $1.4 \times 10^{-3}$ *M*

pH = 2.85

b. HCl is a strong acid and completely ionized so

$[H^+]$ = $3.0 \times 10^{-5}$ *M*

pH = 4.52

c. $HNO_3$ is a strong acid and completely ionized so

$[H^+] = 5.0 \times 10^{-2}$ *M*

pH = 1.30

d. HCl is a strong acid and completely ionized so

$[H^+] = 0.0010$ *M*

pH = 3.00

## 17.6 Buffered Solutions

60. A buffered solution consists of a mixture of a weak acid and its conjugate base; one example of a buffered solution is a mixture of acetic acid ($CH_3COOH$) and sodium acetate ($NaCH_3COO$).

62. The weak acid component of a buffered solution is capable of reacting with added strong base. For example, using the buffered solution given as an example in Question 60, acetic acid would consume added sodium hydroxide as follows:

$CH_3COOH(aq) + NaOH(aq) \rightarrow NaCH_3COO(aq) + H_2O(l)$

Acetic acid *neutralizes* the added NaOH and prevents it from having much effect on the overall pH of the solution.

64. a. added HCl: $NaF(aq) + HCl(aq) \rightarrow HF + NaCl(aq)$

added NaOH: $HF + NaOH(aq) \rightarrow NaF(aq) + H_2O(l)$

b. added HCl: $KCN(aq) + HCl(aq) \rightarrow HCN(aq) + KCl(aq)$

added NaOH: $HCN(aq) + NaOH(aq) \rightarrow NaCN(aq) + H_2O(l)$

c. not buffered

d. not buffered

## Additional Problems

66. a. NaOH is completely ionized, so $[OH^-] = 0.10$ *M*

$pOH = -\log[0.10] = 1.00$

$pH = 14 - 1.00 = 13.00$

b. KOH is completely ionized, so $[OH^-] = 2.0 \times 10^{-4}$ *M*

$pOH = -\log[2.0 \times 10^{-4}] = 3.70$

$pH = 14 - 3.70 = 10.30$

c. CsOH is completely ionized, so $[OH^-] = 6.2 \times 10^{-3}$ *M*

$pOH = -\log[6.2 \times 10^{-3}] = 2.21$

$pH = 14 - 2.21 = 11.79$

d. NaOH is completely ionized, so $[OH^-] = 0.0001$ *M*

$pOH = -\log[0.0001] = 4.0$

$pH = 14 - 4.0 = 10.0$

68. *a*, *b*, and *d* represent solutions; *c* represents an *acidic* solution because there is less hydroxide ion than hydrogen ion.

70. *a*, *c*, and *e*, represent strong acids; *b* and *d* are typical *weak* acids.

72. Ordinarily in calculating the pH of strong acid solutions, the major contribution to the concentration of hydrogen ion present is from the dissolved strong acid; we ordinarily neglect the small amount of hydrogen ion present in such solutions due to the ionization of water.

With $1.0 \times 10^{-7}$ *M* HCl solution, however, the amount of hydrogen ion present due to the ionization of *water* is *comparable* to that present due to the addition of *acid* (HCl) and must be considered in the calculation of pH.

74. accepts

76. base

78. carboxyl (—COOH)

80. $1.0 \times 10^{-14}$

82. larger

84. pH

86. weak acid

## Chapter Eighteen Oxidation-Reduction Reactions/Electrochemistry

### 18.1 Oxidation-Reduction Reactions

2. Oxidation is defined as a *loss of electrons* by an atom, molecule, or ion. Reduction is defined as a *gain of electrons* by such a species.

   $Na \rightarrow Na^+ + e^-$ is an example of an oxidation process.

   $Cl + e^- \rightarrow Cl^-$ is an example of a reduction process.

4. Each of these reactions involves a *metallic* element in the form of the *free* element on one side of the equation; on the other side of the equation, the metallic element is *combined* in an ionic compound. If a metallic element goes from the free metal to the ionic form, the metal is oxidized (loses electrons).

   a. sodium is oxidized, oxygen is reduced

   b. iron is oxidized, hydrogen is reduced

   c. oxygen ($O^{2-}$) is oxidized, aluminum ($Al^{3+}$) is reduced (this reaction is the reverse of the type discussed above)

   d. magnesium is oxidized, nitrogen is reduced

6. Each of these reactions involves a *metallic* element in the form of the *free* element on one side of the equation; on the other side of the equation, the metallic element is *combined* in an ionic compound. If a metallic element goes from the free metal to the ionic form, the metal is oxidized (loses electrons).

   a. zinc is oxidized, nitrogen is reduced

   b. cobalt is oxidized, sulfur is reduced

   c. potassium is oxidized, oxygen is reduced

   d. silver is oxidized, oxygen is reduced

### 18.2 Oxidation States

8. The oxidation state of a *pure element* is *zero*, regardless of whether the element occurs naturally as single atoms or as a molecule.

10. Fluorine is always assigned a negative oxidation state (−1) because all other elements are less electronegative than fluorine. The other halogens are *usually* assigned an oxidation state of −1 in compounds. In interhalogen compounds such as ClF, fluorine is assigned oxidation state −1 (F is more electronegative than Cl), which means that chlorine must be assigned a +1 oxidation state in this instance.

12. Oxidation states represent a bookkeeping method to assign electrons in a molecule or ion. Since a neutral molecule has an overall charge of zero, the sum of the oxidation states in a neutral molecule must be zero. Since an ion has a net charge, the sum of the oxidation states of the atoms in the ion must equal the charge on the ion.

14. The rules for assigning oxidation states are given on page 591. The rule which applies for each element in the following answers is given in parentheses after the element and its oxidation state.

    a. H +1 (Rule 4); N −3 (Rule 6)

    b. C +2 (Rule 6); O −2 (Rule 3)

    c. C +4 (Rule 6); O −2 (Rule 3)

    d. N +3 (Rule 6); F −1 (Rule 5)

16. The rules for assigning oxidation states are given on page 591. The rule which applies for each element in the following answers is given in parentheses after the element and its oxidation state.

    a. C −2 (Rule 6); H +1 (Rule 4)

    b. C −1 (Rule 6); H +1 (Rule 4)

    c. C −3 (Rule 6); H +1 (Rule 4)

    d. C −2 (Rule 6); H +1 (Rule 4); O −2 (Rule 3)

18. The rules for assigning oxidation states are given on page 591. The rule which applies for each element in the following answers is given in parentheses after the element and its oxidation state.

    a. S +4 (Rule 6); O −2 (Rule 3)

    b. S +6 (Rule 6); O −2 (Rule 3)

    c. H +1 (Rule 4); S −2 (Rule 6)

    d. H +1 (Rule 4); S +6 (Rule 6); O −2 (Rule 3)

20. The rules for assigning oxidation states are given on page 591. The rule which applies for each element in the following answers is given in parentheses after the element and its oxidation state.

    a. Cr +3 (Rule 6); Cl −1 (Rule 2)

    b. K +1 (Rule 2); Cr +6 (Rule 6); O −2 (Rule 3)

    c. K +1 (Rule 2); Cr +6 (Rule 6); O −2 (Rule 3)

    d. Cr +2 (Rule 6);C 0 (Rule 7); H +1 (Rule 4); O −2 (Rule 3)

    For $Cr(C_2H_3O_2)_2$, first the oxidation state of carbon in the acetate ion, $C_2H_3O_2^-$, is determined by Rule 7 (the sum of the oxidation states must equal the charge on the ion), then the oxidation state of Cr may be determined by Rule 6).

22. a. Bi +3 (Rule 7); O −2 (Rule 3)
    b. P +5 (Rule 7); O −2 (Rule 3)
    c. N +3 (Rule 7); O −2 (Rule 3)
    d. Hg +1 (Rule 7)

## 18.3 Oxidation-Reduction Reactions Between Nonmetals

24. Electrons are negative; when an atom gains electrons, it gains one negative charge for each electron gained. For example, in the reduction reaction $Cl + e^- \rightarrow Cl^-$, the oxidation state of chlorine decreases from 0 to −1 as the electron is gained.

26. An oxidizing agent *causes* another species to be oxidized (to lose electrons). In order to make another species lose electrons, the oxidizing agent must be capable of gaining the electrons; an oxidizing agent is itself reduced. On the contrary, a reducing agent is itself oxidized.

28. An oxidizing agent oxidizes another species by gaining the electrons lost by the other species; therefore, an oxidizing agent itself decreases in oxidation state. A reducing agent increases its oxidation state when acting on another atom or molecule.

30. a. $C(s) + O_2(g) \rightarrow CO_2(g)$

    C 0   O 0   C +4

    O −2

    carbon is oxidized; oxygen is reduced

    b. $2CO(g) + O_2(g) \rightarrow 2CO_2(g)$

    C +2   O 0   C +4

    O −2   O −2

    carbon (of CO) is oxidized; oxygen (of $O_2$) is reduced

    c. $CH_4(g) + 2O_2(g) \rightarrow CO_2(g) + 2H_2O(g)$

    C −4   O 0   C +4   H +1

    H +1   O −2   O −2

    carbon (of $CH_4$) is oxidized; oxygen (of $O_2$) is reduced

d. $C_2H_2(g) + H_2(g) \rightarrow C_2H_6(g)$

C −1 H 0 C −3

H +1 H +1

hydrogen (of $H_2$) is oxidized; carbon (of $C_2H_2$) is reduced

32. a. $2B_2O_3(s) + 6Cl_2(g) \rightarrow 4BCl_3(l) + 3O_2(g)$

B +3 Cl 0 B +3 O 0

O −2 Cl −1

oxygen is oxidized; chlorine is reduced

b. $GeH_4(g) + O_2(g) \rightarrow Ge(s) + 2H_2O(g)$

Ge −4 O 0 Ge 0 H +1

H +1 O −2

germanium is oxidized; oxygen is reduced

c. $C_2H_4(g) + Cl_2(g) \rightarrow C_2H_4Cl_2(l)$

C −2 Cl 0 C −1

H +1 H +1

Cl −1

carbon is oxidized; chlorine is reduced

d. $O_2(g) + 2F_2(g) \rightarrow 2OF_2(g)$

O 0 F 0 O +2

F −1

oxygen is oxidized; fluorine is reduced

34. Iron is reduced [+3 in $Fe_2O_3(s)$, 0 in $Fe(l)$]; carbon is oxidized [+2 in $CO(g)$, +4 in $CO_2(g)$]. $Fe_2O_3(s)$ is the oxidizing agent; $CO(g)$ is the reducing agent.

36. Chlorine is reduced [0 in $Cl_2(g)$, −1 in $NaCl(s)$]; Bromine is oxidized [−1 in $NaBr(aq)$, 0 in $Br_2(l)$]. $Cl_2(g)$ is the oxidizing agent; $NaBr(aq)$ is the reducing agent.

## 18.4 Balancing Oxidation-Reduction Reactions by the Half-Reaction Method

38. Oxidation-reduction reactions are often more complicated than "regular" reactions; frequently the coefficients necessary to balance the number of electrons transferred come out to be large numbers.

40. Under ordinary conditions it is impossible to have "free" electrons that are not part of some atom, ion, or molecule. For this reason, the total number of electrons lost by the species being oxidized must equal the total number of electrons gained by the species being reduced.

42. a. $I^-(aq) \rightarrow I_2(s)$

Balance iodine: $\mathbf{2}I^-(aq) \rightarrow I_2(s)$

Balance charge: $2I^-(aq) \rightarrow I_2(s) + \mathbf{2e^-}$

Balanced half reaction: $2I^-(aq) \rightarrow I_2(s) + 2e^-$

b. $O_2(g) \rightarrow O^{2-}(s)$

Balance oxygen: $O_2(g) \rightarrow \mathbf{2}O^{2-}(s)$

Balance charge: $O_2(g) + \mathbf{4e^-} \rightarrow 2O^{2-}(s)$

Balanced half reaction: $O_2(g) + 4e^- \rightarrow 2O^{2-}(s)$

c. $P_4(s) \rightarrow P^{3-}(s)$

Balance phosphorus: $P_4(s) \rightarrow \mathbf{4}P^{3-}(s)$

Balance charge: $P_4(s) + \mathbf{12e^-} \rightarrow 4P^{3-}(s)$

Balanced half reaction $P_4(s) + 12e^- \rightarrow 4P^{3-}(s)$

d. $Cl_2(g) \rightarrow Cl^-(aq)$

Balance chlorine: $Cl_2(g) \rightarrow \mathbf{2}Cl^-(aq)$

Balance charge: $Cl_2(g) + \mathbf{2e^-} \rightarrow 2Cl^-(aq)$

Balanced half reaction: $Cl_2(g) + 2e^- \rightarrow 2Cl^-(aq)$

44. a. $SiO_2(s) \rightarrow Si(s)$

Balance oxygen: $SiO_2(s) \rightarrow Si(s) + \mathbf{2H_2O}$

Balance hydrogen: $SiO_2(s) + \mathbf{4H^+}(aq) \rightarrow Si(s) + 2H_2O(l)$

Balance charge: $SiO_2(s) + 4H^+(aq) + \mathbf{4e^-} \rightarrow Si(s) + 2H_2O(l)$

Balanced half reaction: $SiO_2(s) + 4H^+(aq) + 4e^- \rightarrow Si(s) + 2H_2O(l)$

b. $S(s) \rightarrow H_2S(g)$

Balance hydrogen: $S(s) + \mathbf{2H^+}(aq) \rightarrow H_2S(g)$

Balance charge: $S(s) + 2H^+(aq) + \mathbf{2e^-} \rightarrow H_2S(g)$

Balanced half reaction: $S(s) + 2H^+(aq) + 2e^- \rightarrow H_2S(g)$

c. $NO_3^-(aq) \rightarrow HNO_2(aq)$

Balance oxygen: $NO_3^-(aq) \rightarrow HNO_2(aq) + \mathbf{H_2O}(l)$

Balance hydrogen: $NO_3^-(aq) + \mathbf{3H^+}(aq) \rightarrow HNO_2(aq) + H_2O(l)$

Balance charge: $NO_3^-(aq) + 3H^+(aq) + \mathbf{2e^-} \rightarrow HNO_2(aq) + H_2O(l)$

Balanced half reaction: $NO_3^-(aq) + 3H^+(aq) + 2e^- \rightarrow HNO_2(aq) + H_2O(l)$

d. $NO_3^-(aq) \rightarrow NO(g)$

Balance oxygen: $NO_3^-(aq) \rightarrow NO(g) + \mathbf{2H_2O}(l)$

Balance hydrogen: $NO_3^-(aq) + \mathbf{4H^+}(aq) \rightarrow NO(g) + 2H_2O(l)$

Balance charge: $NO_3^-(aq) + 4H^+(aq) + \mathbf{3e^-} \rightarrow NO(g) + 2H_2O(l)$

Balanced half reaction: $NO_3^-(aq) + 4H^+(aq) + 3e^- \rightarrow NO(g) + 2H_2O(l)$

46. For simplicity, the physical states of the substances have been omitted until the final balanced equation is given.

a. $I^-(aq) + MnO_4^-(aq) \rightarrow I_2(aq) + Mn^{2+}(aq)$

$I^- \rightarrow I_2$

Balance iodine: $\mathbf{2}I^- \rightarrow I_2$

Balance charge: $2I^- \rightarrow I_2 + \mathbf{2e^-}$

$MnO_4^- \rightarrow Mn^{2+}$

Balance oxygen: $MnO_4^- \rightarrow Mn^{2+} + \mathbf{4H_2O}$

Balance hydrogen: $\mathbf{8H^+} + MnO_4^- \rightarrow Mn^{2+} + 4H_2O$

Balance charge: $8H^+ + MnO_4^- + \mathbf{5e^-} \rightarrow Mn^{2+} + 4H_2O$

Combine the half reactions: $5 \times (2I^- \rightarrow I_2 + 2e^-)$

$2 \times (8H^+ + MnO_4^- + 5e^- \rightarrow Mn^{2+} + 4H_2O)$

$16H^+(aq) + 2MnO_4^-(aq) + 10I^-(aq) \rightarrow 2Mn^{2+}(aq) + 4H_2O(l) + 5I_2(aq)$

b. $S_2O_8^{2-} + Cr^{3+} \rightarrow SO_4^{2-} + Cr_2O_7^{2-}$

$S_2O_8^{2-} \rightarrow SO_4^{2-}$

Balance sulfur: $S_2O_8^{2-} \rightarrow \mathbf{2}SO_4^{2-}$

Balance charge: $S_2O_8^{2-} + \mathbf{2e^-} \rightarrow 2SO_4^{2-}$

$Cr^{3+} \rightarrow Cr_2O_7^{2-}$

Balance chromium: $\mathbf{2}Cr^{3+} \rightarrow Cr_2O_7^{2-}$

Balance oxygen: $\mathbf{7H_2O} + 2Cr^{3+} \rightarrow Cr_2O_7^{2-}$

Balance hydrogen: $7H_2O + 2Cr^{3+} \rightarrow Cr_2O_7^{2-} + \mathbf{14H^+}$

Balance charge: $7H_2O + 2Cr^{3+} \rightarrow Cr_2O_7^{2-} + 14H^+ + \mathbf{6e^-}$

Combine the half reactions: $3 \times (S_2O_8^{2-} + 2e^- \rightarrow 2SO_4^{2-})$

$7H_2O + 2Cr^{3+} \rightarrow Cr_2O_7^{2-} + 14H^+ + 6e^-$

$$7H_2O(l) + 2Cr^{3+}(aq) + 3S_2O_8^{2-}(aq) \rightarrow Cr_2O_7^{2-}(aq) + 14H^+(aq) + 6SO_4^{2-}(aq)$$

c. $BiO_3^- + Mn^{2+} \rightarrow Bi^{3+} + MnO_4^-$

$BiO_3^- \rightarrow Bi^{3+}$

Balance oxygen: $BiO_3^- \rightarrow Bi^{3+} + \mathbf{3H_2O}$

Balance hydrogen: $\mathbf{6H^+} + BiO_3^- \rightarrow Bi^{3+} + 3H_2O$

Balance charge: $6H^+ + BiO_3^- + \mathbf{2e^-} \rightarrow Bi^{3+} + 3H_2O$

$Mn^{2+} \rightarrow MnO_4^-$

Balance oxygen: $\mathbf{4H_2O} + Mn^{2+} \rightarrow MnO_4^-$

Balance hydrogen: $4H_2O + Mn^{2+} \rightarrow MnO_4^- + \mathbf{8H^+}$

Balance charge: $4H_2O + Mn^{2+} \rightarrow MnO_4^- + 8H^+ + \mathbf{5e^-}$

Combine the half reactions: $5 \times (6H^+ + BiO_3^- + 2e^- \rightarrow Bi^{3+} + 3H_2O)$

$2 \times (4H_2O + Mn^{2+} \rightarrow MnO_4^- + 8H^+ + 5e^-)$

$$2Mn^{2+}(aq) + 14H^+(aq) + 5BiO_3^-(aq) \rightarrow 2MnO_4^-(aq) + 5Bi^{3+}(aq) + 7H_2O(l)$$

48. For simplicity, the physical states of the substances have been omitted until the final balanced equation is given.

For the reduction of the permanganate ion, $MnO_4^-$, in acid solution, the half reaction is always the *same*:

$MnO_4^- \rightarrow Mn^{2+}$

Balance oxygen: $MnO_4^- \rightarrow Mn^{2+} + \mathbf{4H_2O}$

Balance hydrogen: $\mathbf{8H^+} + MnO_4^- \rightarrow Mn^{2+} + 4H_2O$

Balance charge: $8H^+ + MnO_4^- + \mathbf{5e^-} \rightarrow Mn^{2+} + 4H_2O$

a. $C_2O_4^{2-} \rightarrow CO_2$

Balance carbon: $C_2O_4^{2-} \rightarrow 2CO_2$

Balance charge: $C_2O_4^{2-} \rightarrow 2CO_2 + 2e^-$

Combine half reactions: $5 \times (C_2O_4^{2-} \rightarrow 2CO_2 + 2e^-)$

$2 \times (8H^+ + MnO_4^- + 5e^- \rightarrow Mn^{2+} + 4H_2O)$

$$16H^+(aq) + 2MnO_4^-(aq) + 5C_2O_4^{2-}(aq) \rightarrow 2Mn^{2+}(aq) + 8H_2O(l) + 10CO_2(g)$$

b. $Fe^{2+} \rightarrow Fe^{3+}$

Balance charge: $Fe^{2+} \rightarrow Fe^{3+} + e^-$

Combine half reactions: $5 \times (Fe^{2+} \rightarrow Fe^{3+} + e^-)$

$8H^+ + MnO_4^- + 5e^- \rightarrow Mn^{2+} + 4H_2O$

$$8H^+(aq) + MnO_4^-(aq) + 5Fe^{2+}(aq) \rightarrow Mn^{2+}(aq) + 4H_2O(l) + 5Fe^{3+}(aq)$$

c. $Cl^- \rightarrow Cl_2$

Balance chlorine: $2Cl^- \rightarrow Cl_2$

Balance charge: $2Cl^- \rightarrow Cl_2 + 2e^-$

Combine half reactions: $5 \times (2Cl^- \rightarrow Cl_2 + 2e^-)$

$2 \times (8H^+ + MnO_4^- + 5e^- \rightarrow Mn^{2+} + 4H_2O)$

$$16H^+(aq) + 2MnO_4^-(aq) + 10Cl^-(aq) \rightarrow 2Mn^{2+}(aq) + 8H_2O(l) + 5Cl_2(g)$$

## 18.5 Electrochemistry: An Introduction

50. A salt bridge typically consists of a *U*-shaped tube filled with an inert electrolyte (one involving ions that are not part of the oxidation-reduction reaction). A salt bridge is used to complete the electrical circuit in a cell. Any method which allows transfer of charge without allowing bulk mixing of the solutions may be used (another common method is to set up one half-cell in a porous cup, which is then placed in the beaker containing the second half-cell).

52. In a galvanic cell, the anode is the electrode where oxidation occurs; the cathode is the electrode where reduction occurs.

54.

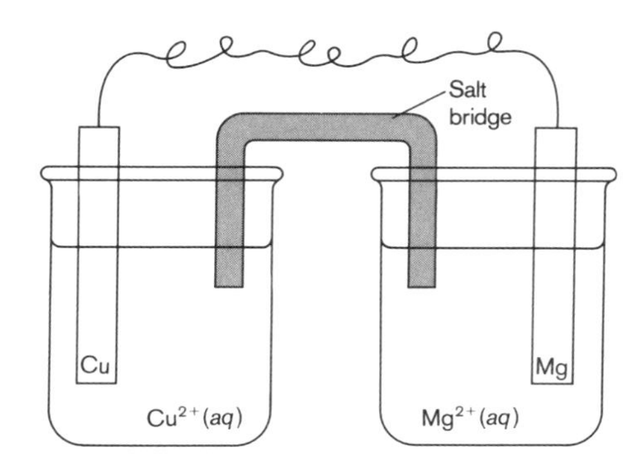


$Cu^{2+}(aq)$ ion is reduced; $Mg(s)$ is oxidized.

The reaction at the anode is $Mg(s) \rightarrow Mg^{2+}(aq) + 2e^-$.

The reaction at the cathode is $Cu^{2+}(aq) + 2e^- \rightarrow Cu(s)$.

## 18.6 Batteries

56. Both normal and alkaline cells contain zinc as one electrode; zinc corrodes more slowly under alkaline conditions than in the highly acidic environment of a normal dry cell.

anode: $Zn(s) + 2OH^-(aq) \rightarrow ZnO(s) + H_2O(l) + 2e^-$

cathode: $2MnO_2(s) + H_2O(l) + 2e^- \rightarrow Mn_2O_3(s) + 2OH^-(aq)$

## 18.7 Corrosion

58. Aluminum is a very reactive metal when freshly isolated in the pure state. However, on standing for even a relatively short period of time, aluminum metal forms a thin coating of $Al_2O_3$ on its surface from reaction with atmospheric oxygen. This coating of $Al_2O_3$ is much less reactive than the metal and serves to protect the surface of the metal from further attack.

60. In cathodic protection of steel tanks and pipes, a more reactive metal than iron is connected to the item to be protected. The active metal is then preferentially oxidized rather than the iron of the tank or pipe.

## 18.8 Electrolysis

62. The main recharging reaction for the lead storage battery is

$$2PbSO_4(s) + 2H_2O(l) \rightarrow Pb(s) + PbO_2(s) + 2H^+(aq) + 2HSO_4^-(aq)$$

A major side reaction is the electrolysis of water

$2H_2O(l) \rightarrow 2H_2(g) + O_2(g)$

resulting in the production of an explosive mixture of hydrogen and oxygen, which accounts for many accidents during the recharging of such batteries.

64. Electrolysis is applied in electroplating by making the item to be plated the cathode in a cell containing a solution of ions of the desired plating metal.

## Additional Problems

66. loss, oxidation state

68. electronegative

70. zero

72. lose

74. separated

76. anode

78. electrolysis

80. hydrogen, oxygen

82. oxidation

84. a. $4Fe(s) + 3O_2(g) \rightarrow 2Fe_2O_3(s)$
iron is oxidized, oxygen is reduced

b. $2Al(s) + 3Cl_2(g) \rightarrow 2AlCl_3(s)$
aluminum is oxidized, chlorine is reduced

c. $6Mg(s) + P_4(s) \rightarrow 2Mg_3P_2(s)$
magnesium is oxidized, phosphorus is reduced

86. a. magnesium is oxidized, aluminum (as $Al^{3+}$) is reduced
b. magnesium is oxidized, aluminum (as $Al^{3+}$ in $AlCl_3$) is reduced
c. oxygen is oxidized, chlorine is reduced

88. a. $C_3H_8(g) + 5O_2(g) \rightarrow 3CO_2(g) + 4H_2O(g)$
b. $CO(g) + 2H_2(g) \rightarrow CH_3OH(l)$
c. $SnO_2(s) + 2C(s) \rightarrow Sn(s) + 2CO(g)$
d. $C_2H_5OH(l) + 3O_2(g) \rightarrow 2CO_2(g) + 3H_2O(g)$

# Chapter Nineteen Radioactivity and Nuclear Energy

## 19.1 Radioactive Decay

2. The nucleus consists of protons (mass number 1, charge 1+) and neutrons (mass number 1, charge 0).

4. Isotopes represent atoms of the same element with different mass numbers. On a nuclear basis, isotopes differ in the number of neutrons present. For samples with large numbers of atoms, isotopes have the same chemical properties, but may have slightly different physical properties.

6. A nucleus is termed radioactive if it spontaneously decomposes, forming another nucleus and producing one or more subatomic particles.

8. Alpha particle: charge 2+, mass number 4, symbol ${}^{4}_{2}He$

10. When a nucleus produces a beta particle, the atomic number of the parent nucleus is *increased* by *one* unit.

12. Gamma rays are high energy photons of electromagnetic radiation. Gamma rays are not normally considered to be particles. When a nucleus produces only gamma radiation, the atomic number and mass number of the nucleus do not change.

14. Electron capture occurs when one of the inner orbital electrons is pulled into and becomes part of the nucleus.

16. ${}^{35}_{17}Cl$ and ${}^{37}_{17}Cl$

18. ${}^{27}_{13}Al$: 13 protons, 14 neutrons

    ${}^{28}_{13}Al$: 13 protons, 15 neutrons

    ${}^{29}_{13}Al$: 13 protons, 16 neutrons

20. a. ${}^{0}_{-1}e$ or ${}^{0}_{-1}\beta$

    b. ${}^{0}_{+1}e$ or ${}^{0}_{+1}\beta$

    c. ${}^{0}_{0}\gamma$

22. a. ${}^{4}_{2}He$

b. $^{4}_{2}He$

c. $^{1}_{0}n$

24. a. $^{1}_{0}n$

b. $^{201}_{79}Au$

c. $^{0}_{-1}e$

26. a. $^{59}_{26}Fe \rightarrow ^{59}_{27}Co + ^{0}_{-1}e$

b. $^{24}_{11}Na \rightarrow ^{24}_{12}Mg + ^{0}_{-1}e$

c. $^{198}_{79}Au \rightarrow ^{198}_{80}Hg + ^{0}_{-1}e$

d. $^{47}_{20}Ca \rightarrow ^{47}_{21}Sc + ^{0}_{-1}e$

28. a. $^{226}_{88}Ra \rightarrow ^{222}_{86}Rn + ^{4}_{2}He$

b. $^{222}_{86}Rn \rightarrow ^{218}_{84}Po + ^{4}_{2}He$

c. $^{239}_{94}Pu \rightarrow ^{235}_{92}U + ^{4}_{2}He$

d. $^{8}_{4}Be \rightarrow ^{4}_{2}He + ^{4}_{2}He$

## 19.2 Nuclear Transformations

30. There is often considerable repulsion between the target nucleus and the particles being used for bombardment (especially if the bombarding particle is positively charged like the target nucleus). Using accelerators to greatly speed up the bombarding particles can overcome this repulsion.

32. $^{14}_{7}N + ^{4}_{2}He \rightarrow ^{17}_{8}O + ^{1}_{1}H$

## 19.3 Detection of Radioactivity and the Concept of Half-Life

34. The half-life of a nucleus is the time required for one-half of the original sample of nuclei to decay. A given isotope of an element always has the same half-life, although different isotopes of the same element may have greatly different half-lives. Nuclei of different elements have different half-lives.

36. $^{226}_{88}Ra$ is the most stable (longest half-life)

    $^{224}_{88}Ra$ is the "hottest" (shortest half-life)

38. highest activity → lowest activity

    $^{87}Sr > ^{99}Tc > ^{24}Na > ^{99}Mo > ^{133}Xe > ^{131}I > ^{32}P > ^{51}Cr > ^{59}Fe$

40. For a decay of 10 $\mu g$ to 1/1000 of this amount, we want to know when the amount of remaining amount of $^{131}I$ is on the order of 0.01 $\mu g$.

| time, days | 0 | 8 | 16 | 24 | 32 | 40 | 48 | 56 | 64 | 72 | 80 |
|---|---|---|---|---|---|---|---|---|---|---|---|
| mass, $\mu g$ | 10 | 5 | 2.5 | 1.25 | 0.625 | 0.313 | 0.156 | 0.078 | 0.039 | 0.020 | 0.01 |

Approximately 80 days are required.

42. For simplicity, assume 100 $\mu g$ was administered initially; 2 days = 48 hours

| time, hours | 0 | 6 | 12 | 18 | 24 | 30 | 36 | 42 | 48 |
|---|---|---|---|---|---|---|---|---|---|
| mass, $\mu g$ | 100 | 50 | 25 | 12.5 | 6.25 | 3.13 | 1.56 | 0.78 | 0.39 |

For an administered dose of 100 $\mu g$, 0.39 $\mu g$ remains after 2 days. The fraction remaining is 0.39/100 = 0.0039; on a percentage basis, less than 0.4 % of the original radioisotope remains.

## 19.4 Dating by Radioactivity

44. Carbon-14 is produced in the upper atmosphere by the bombardment of ordinary nitrogen with neutrons from space:

$$^{14}_{7}N + ^{1}_{0}n \rightarrow ^{14}_{6}C + ^{1}_{1}H$$

46. We assume that the concentration of C-14 in the atmosphere is effectively constant. A living organism is constantly replenishing C-14 either through the processes of metabolism (sugars ingested in foods contain C-14), or photosynthesis (carbon dioxide contains C-14). When a plant dies, it no longer replenishes itself with C-14 from the atmosphere, and as the C-14 undergoes radioactive decay, its amount decreases with time.

## 19.5 Medical Applications of Radioactivity

48. $^{131}I$ is used in the diagnosis and treatment of thyroid cancer and other disfunctions of the thyroid gland. The thyroid gland is the only place in the human body which uses and stores iodine. I-131 that is administered concentrates in the thyroid, and can be used to cause an image on a scanner or x-ray film, or in higher doses, to selectively kill cancer cells in the thyroid. $^{201}Tl$ concentrates in healthy muscle cells when administered, and can be used to detect and assess damage to heart muscles after a heart attack: the damaged muscles show a lower uptake of Tl-201 than normal muscles.

## 19.6 Nuclear Energy

50. Combining two light nuclei to form a heavier, more stable nucleus is called nuclear *fusion*. Splitting a heavy nucleus into nuclei with smaller mass numbers is called nuclear *fission*.

## 19.7 Nuclear Fission

52. $^{1}_{0}n + ^{235}_{92}U \rightarrow ^{142}_{56}Ba + ^{91}_{36}Kr + 3\,^{1}_{0}n$

54. A critical mass of a fissionable material is the amount needed to provide a high enough internal neutron flux to sustain the chain reaction (enough neutrons are produced to cause the continuous fission of further material). A sample with less than a critical mass is still radioactive, but cannot sustain a chain reaction.

## 19.8 Nuclear Reactors

56. An actual nuclear explosion, of the type produced by a nuclear weapon, cannot occur in a nuclear reactor because the concentration of the fissionable materials is not sufficient to form a supercritical mass. However, since many reactors are cooled by water, a chemical explosion is possible which could scatter the radioactive material used in the reactor.

58. Breeder reactors are set up to convert non-fissionable $^{238}U$ into fissionable $^{239}Pu$. The material used for fission in a breeder reactor is a combination of U-235 (which undergoes fission in a chain reaction) and the more common U-238 isotope. Excess neutrons from the U-235 fission are absorbed by the U-238 converting it to the fissionable plutonium isotope Pu-239. Although Pu-239 is fissionable, its chemical and physical properties make it very difficult and expensive to handle and process.

## 19.9 Nuclear Fusion

60. In one type of fusion reactor, two $^{2}_{1}H$ atoms are fused to produce $^{4}_{2}He$. Because the hydrogen nuclei are positively charged, extremely high energies (temperatures of 40 million K) are needed to overcome the repulsion between the nuclei as they are shot into each other.

62 "Cold" fusion implies the fusing of hydrogen atoms into helium at or near ordinary laboratory temperatures. This method seems preferable since it should be cheaper and more easily controllable. Certain metals (palladium and platinum, for example) have been known for a long time to be capable of dissolving large quantities of hydrogen or deuterium (there is a great deal of space between the metal atoms in the metal's crystal structure). It is thought that if deuterium can be concentrated sufficiently inside the metal's lattice, then deuterium atoms may come close enough together to fuse. Recent experiments have suggested that this might be possible through electrolysis of heavy water using electrodes made of such metals, although these experiments remain unconfirmed at the time this is written.

## 19.10 Effects of Radiation

64. Somatic damage is damage directly to the organism itself, causing nearly immediate sickness or death to the organism. Genetic damage is damage to the genetic machinery of the organism, which will be manifested in future generations of offspring.

66. Gamma rays penetrate long distances, but seldom cause ionization of biological molecules. Alpha particles, because they are much heavier although less penetrating, are very effective at ionizing biological molecules and leave a dense trail of damage in the organism. Alpha particles can be ingested or breathed into the body where the damage will be more acute.

68. The exposure limits given in Table 19.5 as causing no detectable clinical effect are 0–25 rem. The total yearly exposures from natural and human-induced radioactive sources are estimated in Table 19.6 as less than 200 *milli*rem (0.2 rem), which is well within the acceptable limits.

## Additional Problems

70. radioactive

72. mass

74. neutron, proton

76. radioactive decay

78. mass number

80. transuranium

82. half-life

84. radiotracers

86. chain

88. breeder

90. $\frac{\$10}{1\ \mu g} \times \frac{10^6\ \mu g}{1\ g} \times \frac{454\ g}{1\ lb} = \$4.5 \times 10^9$ (4.5 billion dollars!)

92. $\frac{2.1 \times 10^{13}\ J}{1\ mol} \times \frac{1\ mol}{6.02 \times 10^{23}\ atoms} = 3.5 \times 10^{-11}\ J/atom$

$\frac{2.1 \times 10^{13}\ J}{1\ mol} \times \frac{1\ mol}{235\ g} = 8.9 \times 10^{10}\ J/g$

94. Despite the fact that nuclear waste has been generated for over 40 years, no permanent disposal plan has been implemented as yet. One proposal to dispose of such waste calls for the waste to be sealed in blocks of glass, which in turn are sealed in corrosion-proof metal drums, which would then be buried in deep, stable rock formations away from earthquake and other geologically active zones. In these deep storage areas, it is hoped that the waste could decay safely undisturbed until the radioactivity drops to "safe" levels.

## Chapter Twenty

### 20.1 Carbon Bon

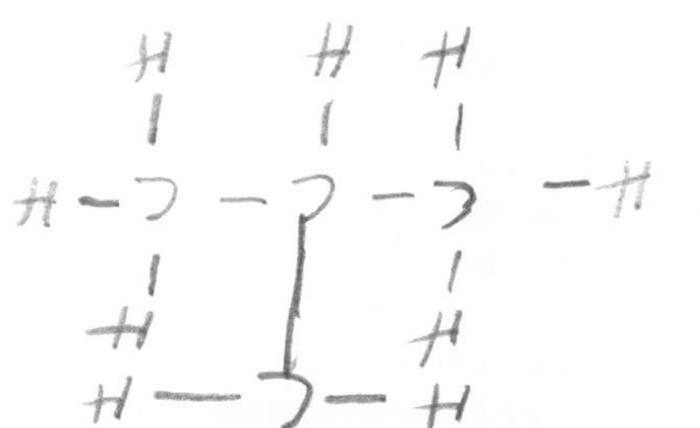

2. A carbon atc ... r atoms.

4. A triple boı ... trons (three pairs of electrons). ... nolecule containing a triple bond

6. Each carbon ... carbon atom has its electron pa

### 20.2 Alkanes

8. A saturated hydrocarbon is one in which all the bonds are *single* bonds, with each carbon atom forming bonds to four other atoms. The saturated chain hydrocarbons are called *alkanes*.

10. A "straight-chain" alkane is not really straight because the electron pairs on the carbon atoms have a tetrahedral orientation, separated by an angle of 109.5°. In order to give a truly straight chain, the angle between electron pairs would have to be 90° and multiples of 90°.

12. a.

```
  H H H H H H H
  | | | | | | |
H-C-C-C-C-C-C-C-H
  | | | | | | |
  H H H H H H H
```

b.

```
  H H H H H H H H H H
  | | | | | | | | | |
H-C-C-C-C-C-C-C-C-C-C-H
  | | | | | | | | | |
  H H H H H H H H H H
```

c.

```
  H H H H H H H H H H H H H H H
  | | | | | | | | | | | | | | |
H-C-C-C-C-C-C-C-C-C-C-C-C-C-C-C-H
  | | | | | | | | | | | | | | |
  H H H H H H H H H H H H H H H
```

14. a.

```
  H H H H H H H H
  | | | | | | | |
H-C-C-C-C-C-C-C-C-H
  | | | | | | | |
  H H H H H H H H
```

octane $CH_3CH_2CH_2CH_2CH_2CH_2CH_2CH_3$

b.

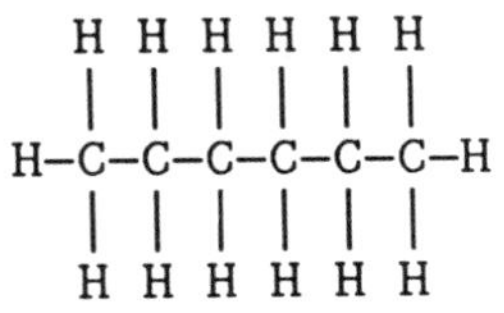

```
  H H H H H H
  | | | | | |
H-C-C-C-C-C-C-H
  | | | | | |
  H H H H H H
```

hexane $CH_3CH_2CH_2CH_2CH_2CH_3$

c.

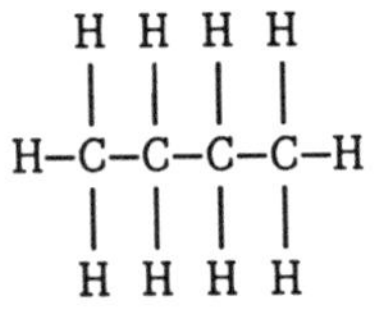

```
  H H H H
  | | | |
H-C-C-C-C-H
  | | | |
  H H H H
```

butane $CH_3CH_2CH_2CH_3$

d.

```
  H H H H H
  | | | | |
H-C-C-C-C-C-H
  | | | | |
  H H H H H
```

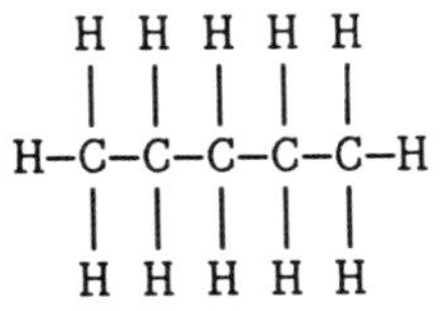

pentane $CH_3CH_2CH_2CH_2CH_3$

## 20.3 Structural Formulas and Isomerism

16. branch

18. The carbon skeletons are

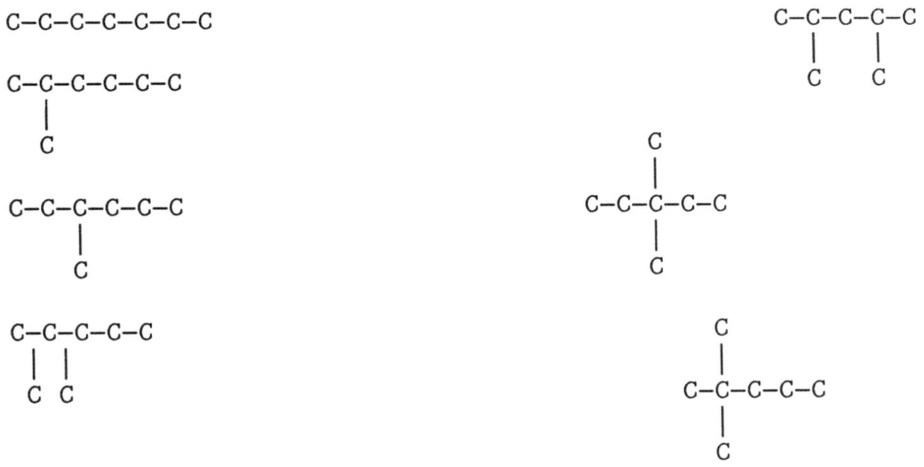

```
C-C-C-C-C-C-C

C-C-C-C-C-C
  |
  C

C-C-C-C-C-C
    |
    C

C-C-C-C-C
  | |
  C C

C-C-C-C-C
  |   |
  C   C

    C
    |
C-C-C-C-C
    |
    C

  C
  |
C-C-C-C-C
  |
  C
```

## 20.4 Naming Alkanes

20. The root name is derived from the number of carbon atoms in the longest continuous chain of carbon atoms.

22. The position of a substituent is indicated by a number which corresponds to the carbon atom in the longest chain to which the substituent is attached.

24. Multiple substituents are listed in alphabetical order, disregarding any prefix.

26. a. 2,3-dimethylbutane

    b. 3,3-diethylpentane

    c. 2,3,3-trimethylhexane

    d. 2,3,4,5,6-pentamethylheptane

28. a.

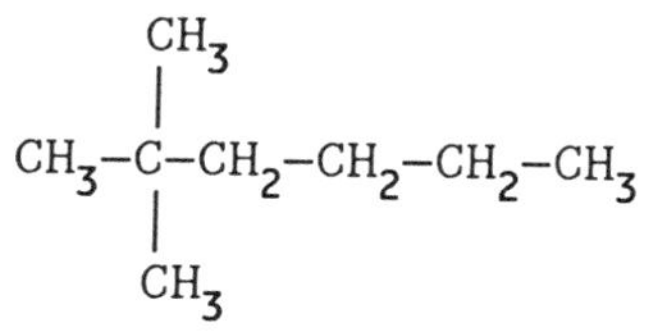

```
         CH3
         |
    CH3–C–CH2–CH2–CH2–CH3
         |
         CH3
```

b.

```
              CH3
              |
    CH3–CH–CH–CH2–CH2–CH3
         |
         CH3
```

c.

```
                 CH3
                 |
    CH3–CH2–C–CH2–CH2–CH3
                 |
                 CH3
```

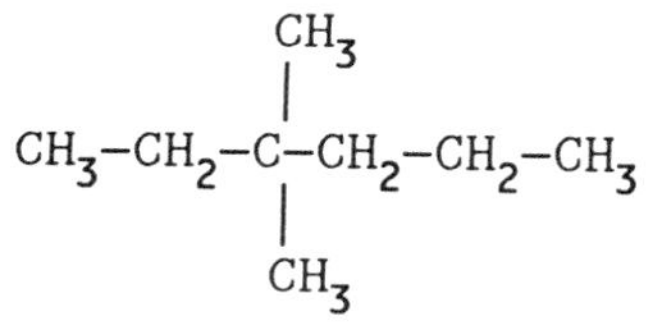

d.

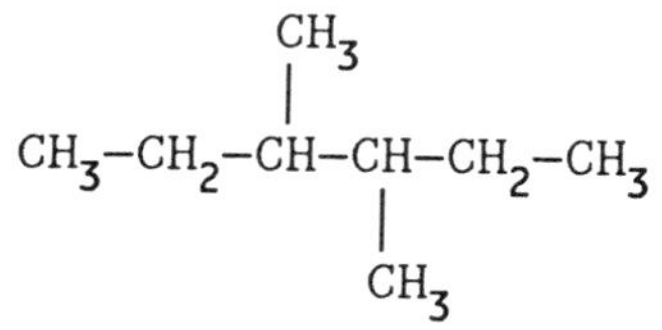

```
                 CH3
                 |
    CH3–CH2–CH–CH–CH2–CH3
                    |
                    CH3
```

e.

```
         CH3
         |
    CH3–CH–CH2–CH–CH2–CH3
                    |
                    CH3
```

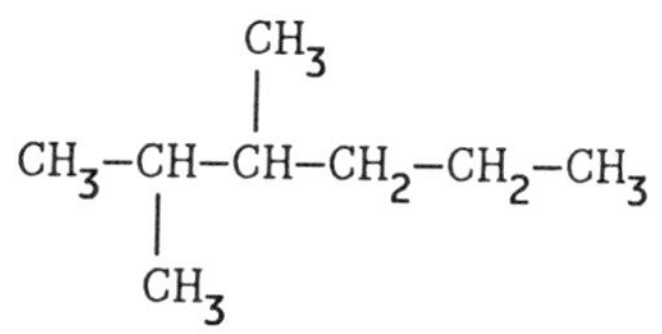

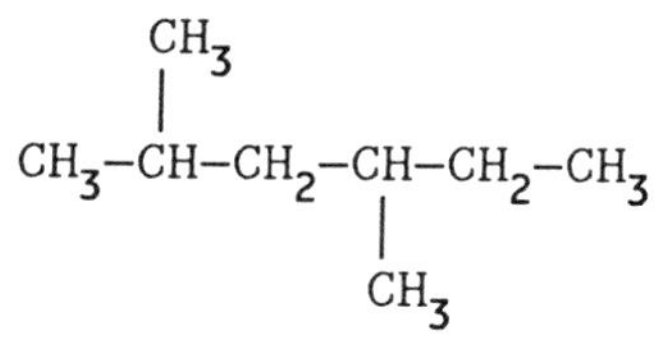

## 20.5 Petroleum

30.

| *Number of C atoms* | *Use* |
|---|---|
| $C_5$–$C_{12}$ | gasoline |
| $C_{10}$–$C_{18}$ | kerosene, jet fuel |
| $C_{15}$–$C_{25}$ | diesel fuel, heating oil, lubrication |
| $C_{25}$– | asphalt |

32. Tetraethyl lead was added to gasolines to prevent "knocking" of high efficiency automobile engines. The use of tetraethyl lead is being phased out because of the danger to the environment of the lead in this substance.

## 20.6 Reactions of Alkanes

34. Combustion represents the vigorous reaction of a hydrocarbon (or other substance) with oxygen. The combustion of alkanes has been made use of as a source of heat and light.

36. Dehydrogenation reactions involve the removal of hydrogen atoms from adjacent carbon atoms in an alkane (or other substance). When two hydrogen atoms are removed, a double bond is created.

38. a. $CH_3Cl(g)$

b. $H_2(g)$

c. $HCl(g)$

## 20.7 Alkenes and Alkynes

40. An alkyne is a hydrocarbon containing a carbon-carbon triple bond. The general formula is $C_nH_{2n-2}$.

42. The location of a double or triple bond in the longest chain of an alkene or alkyne is indicated by giving the *number* of the lowest number carbon atom involved in the bond.

44. hydrogenation

46. a. 2-decene

b. 2-heptene

c. 2-pentyne

48. $CH{\equiv}C{-}CH_2{-}CH_2{-}CH_2{-}CH_2{-}CH_2{-}CH_3$ 1-octyne

$CH_3{-}C{\equiv}C{-}CH_2{-}CH_2{-}CH_2{-}CH_2{-}CH_3$ 2-octyne

$CH_3-CH_2-C \equiv C-CH_2-CH_2-CH_2-CH_3$ 3-octyne

$CH_3-CH_2-CH_2-C \equiv C-CH_2-CH_2-CH_3$ 4-octyne

## 20.8 Aromatic Hydrocarbons

50. For benzene, a *set* of equivalent Lewis structures can be drawn, differing only in the location of the three double bonds in the ring. Experimentally, however, benzene does not demonstrate the chemical properties expected for molecules having *any* double bonds.

## 20.9 Naming Aromatic Compounds

52. When named as a substituent, the benzene ring is called the phenyl group. Two examples are

$CH_2{=}CH-CH(C_6H_5)-CH_3$ $CH_3-CH(C_6H_5)-CH_2-CH_2-CH_2-CH_3$

3-phenyl-1-butene 2-phenylhexane

54. *ortho-* refers to adjacent substituents (1,2-); *meta-* refers to two substituents with one unsubstituted carbon atom between them (1,3-); *para-* refers to two substituents with two unsubstituted carbon atoms between them (1,4-).

56. a. 2-ethylmethylbenzene (2-ethyltoluene)
    b. 1,3,5-tribromobenzene
    c. 2-chloronitrobenzene
    d. naphthalene

## 20.10 Functional Groups

58. a. carboxylic acid
    b. ketone
    c. ester
    d. alcohol (phenol)

## 20.11 Alcohols

60. Primary alcohols have one hydrocarbon fragment (alkyl group) bonded to the carbon atom where the –OH group is attached. Secondary alcohols have two such alkyl groups attached, and tertiary alcohols contain three such alkyl groups. Examples are

ethanol (primary) $CH_3{-}CH_2{-}OH$

2-propanol (secondary)

$$\begin{array}{ccc} CH_3- & CH & -CH_3 \\ & | & \\ & OH & \end{array}$$

2-methyl-2-propanol (tertiary)

$$\begin{array}{ccc} & CH_3 & \\ & | & \\ CH_3- & C & -CH_3 \\ & | & \\ & OH & \end{array}$$

62. a. 2-propanol (secondary)

$$\begin{array}{ccc} CH_3- & CH & -CH_3 \\ & | & \\ & OH & \end{array}$$

b. 2-methyl-2-propanol (tertiary)

$$\begin{array}{ccc} & CH_3 & \\ & | & \\ CH_3- & C & -CH_3 \\ & | & \\ & OH & \end{array}$$

c. 4-isopropyl-2-heptanol (secondary)

$$\begin{array}{ccccccc} & & & CH_3{-}CH{-}CH_3 & & & \\ & & & | & & & \\ CH_3- & CH & -CH_2- & CH & -CH_2- & CH_2- & CH_3 \\ & | & & & & & \\ & OH & & & & & \end{array}$$

d. 2,3-dichloro-1-pentanol (primary)

$$\begin{array}{ccccc} CH_2- & CH- & CH- & CH_2- & CH_3 \\ | & | & | & & \\ OH & Cl & Cl & & \end{array}$$

## 20.12 Properties and Uses of Alcohols

64. $C_6H_{12}O_6 \xrightarrow{\text{yeast}} 2CH_3{-}CH_2{-}OH + CO_2$

The yeast necessary for the fermentation process are killed if the concentration of ethanol is over 13%. More concentrated ethanol solutions are most commonly made by distillation.

66. methanol ($CH_3OH$) - starting material for synthesis of acetic acid and many plastics

ethylene glycol ($CH_2OH-CH_2OH$) - automobile antifreeze

isopropyl alcohol (2-propanol, $CH_3-CH(OH)-CH_3$) - rubbing alcohol

## 20.13 Aldehydes and Ketones

68. Aldehydes and ketones both contain the carbonyl group ($\backslash C=O /$).

Aldehydes and ketones differ in the *location* of the carbonyl function: aldehydes contain the carbonyl group at the end of a hydrocarbon chain (the carbon atom of the carbonyl group is bonded only to at most one other carbon atom); the carbonyl group of ketones represents one of the interior carbon atoms of a chain (the carbon atom of the carbonyl group is bonded to two other carbon atoms).

70. Aldehydes and ketones are produced by the oxidation of primary and secondary alcohols, respectively.

$CH_3-CH_2-OH$ ---oxidation⟶ $CH_3-C(H)=O$

$CH_3-CH(CH_3)-OH$ ---oxidation⟶ $CH_3-C(CH_3)=O$

## 20.14 Naming Aldehydes and Ketones

72. In addition to their systematic names (based on the hydrocarbon root, with the ending *-one*), ketones can also be named by naming the groups attached to either side of the carbonyl carbon as alkyl groups, followed by the word "ketone"

$CH_3-CH_2-C(CH_2-CH_3)=O$
3-pentanone

diethyl ketone

$CH_3-C(CH_2-CH_2-CH_3)=O$
2-pentanone

methyl propyl ketone

74. a. 3-methylpentanal

$CH_3-CH_2-CH(CH_3)-CH_2-C(H)=O$

b. 3-methyl-2-pentanone

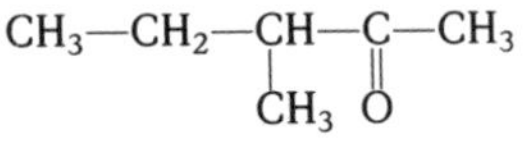

$$CH_3-CH_2-\underset{CH_3}{\underset{|}{CH}}-\underset{O}{\underset{\|}{C}}-CH_3$$

c. methyl phenyl ketone

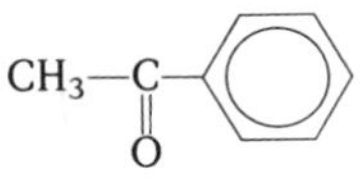

$$CH_3-\underset{O}{\underset{\|}{C}}-C_6H_5$$

d. 2-hydroxybutanal

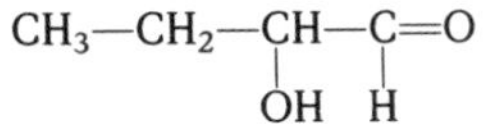

$$CH_3-CH_2-\underset{OH}{\underset{|}{CH}}-\underset{H}{\underset{|}{C}}=O$$

e. propanal

$$CH_3-CH_2-\underset{H}{\underset{|}{C}}=O$$

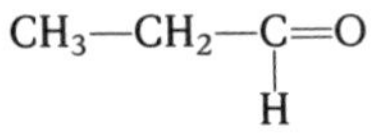

## 20.15 Carboxylic Acids and Esters

76. Carboxylic acids are typically *weak* acids.

$$CH_3-CH_2-COOH(aq) \rightleftharpoons H^+(aq) + CH_3-CH_2-COO^-(aq)$$

78. Carboxylic acids are synthesized from the corresponding primary alcohol by strong oxidation with a reagent such as potassium permanganate

$$CH_3-CH_2-CH_2-OH \xrightarrow{KMnO_4} CH_3-CH_2-COOH$$

The synthesis of carboxylic acids from alcohols is an oxidation/reduction reaction.

80. Acetylsalicylic acid is synthesized from salicylic acid (behaving as an alcohol through its –OH group) and acetic acid.

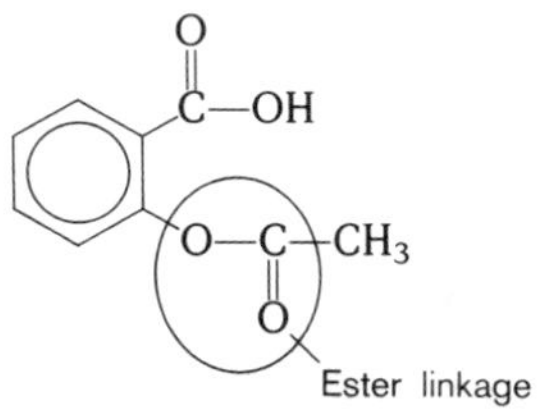

$$C_6H_4(-\overset{O}{\overset{\|}{C}}-OH)(-O-\underset{O}{\underset{\|}{C}}-CH_3)$$

Ester linkage

82. a. $CH_3-CH(CH_3)-CH_2-COOH$

b. (2-chlorobenzoic acid: benzene ring with $-\overset{O}{\overset{\|}{C}}-OH$ and $-Cl$ on adjacent carbons)

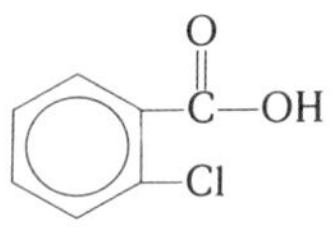

c. $CH_3-CH_2-CH_2-CH_2-CH_2-COOH$

d. $CH_3-COOH$

## 20.16 Polymers

84. In addition polymerization, the monomer units simply add together to form the polymer, with no other products. Polyethylene and polytetrafluoroethylene (Teflon) are common examples.

86. A polyester is formed from the reaction of a dialcohol (two –OH groups) with a diacid (two –COOH groups). One –OH group of the alcohol forms an *ester linkage* with one of the –COOH groups of the acid. Since the resulting dimer still possesses an –OH and a –COOH group, the dimer can undergo further esterification reactions. Dacron is a common polyester.

88. nylon

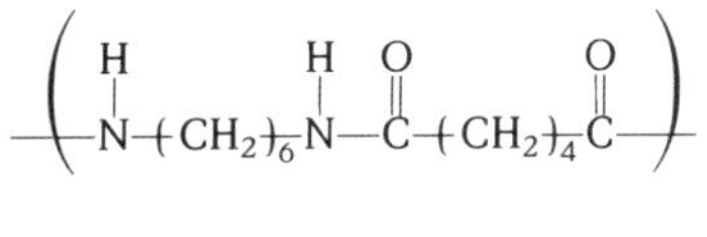

$$\left( \begin{matrix} H \\ | \\ -N \end{matrix} -(CH_2)_6- \begin{matrix} H \\ | \\ N \end{matrix} - \begin{matrix} O \\ \| \\ C \end{matrix} -(CH_2)_4- \begin{matrix} O \\ \| \\ C- \end{matrix} \right)$$

dacron

$$\left( -O-CH_2-CH_2-O-\overset{O}{\overset{\|}{C}}-C_6H_4-\overset{O}{\overset{\|}{C}}- \right)$$

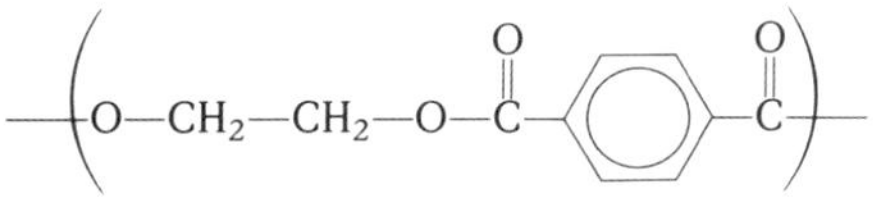

## Additional Problems

90. saturated

92. straight-chain or normal

94. -*ane*

96. number

98. anti-knock

100. substitution

102. hydrogenation

104. functional

106. carbon monoxide

108. carbonyl

110. carboxyl

112. addition

114. The carbon skeletons are

```
C–C–C–C–C            C–C–C–C
                       |
                       C
```

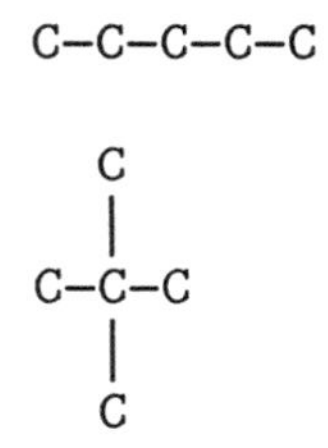

116. a. 2-chlorobutane
     b. 1,2-dibromoethane
     c. triiodomethane (common name: iodoform)
     d. 2,3,4-trichloropentane
     e. 2,2-dichloro-4-isopropylheptane

118. a.

$$CH_3{-}CH_2{-}\underset{\displaystyle |\atop \displaystyle Cl}{C}{=}CH{-}CH_2{-}CH_3$$

b.

$$\underset{\displaystyle |\atop \displaystyle Br}{CH_2}{-}\underset{\displaystyle |\atop \displaystyle Br}{CH_2}$$

c.
```
Cl Cl
|  |
CH-CH
|  |
Cl Cl
```

d.
```
CH≡C-CH-CH2-CH2-CH3
     |
     Br
```

e.
```
       CH3  CH3
       |    |
CH2-C----C----CH-CH2-CH2-CH2-CH3
|      |    |    |
Cl     Cl   CH3  CH3
```

120. primary

$CH_3—CH_2—CH_2—CH_2—CH_2—CH_2—OH$

secondary
```
CH3—CH—CH2—CH2—CH2—CH3
    |
    OH
```

tertiary
```
    CH3
    |
CH3—C—CH2—CH2—CH3
    |
    OH
```

122.
```
CH2-CH-CH-CH-CH-C=O
|   |  |  |  |  |
OH  OH OH OH OH H
```

124. a.
```
       O
       ‖
(C6H5)—C—(C6H5)
```

b.
```
    CH3
    |
CH3—C—CH3
    |
    OH
```

c.

$$\begin{array}{l} CH_3\ \ O \\ \ |\qquad \| \\ CH-C-CH_3 \\ \ | \\ CH_3 \end{array}$$

d.

$$\begin{array}{l} CH_2-CH_2 \\ \ |\qquad\ \ | \\ OH\quad OH \end{array}$$

e.

$$CH_3-CH_2-C{\overset{\displaystyle O}{\diagup\!\!\diagup}} \atop \qquad\qquad\quad \diagdown O-CH_3$$

126.

$$\begin{array}{l} HO-C=O \\ \qquad\ \ | \\ CH_3-CH-N-C=O + H_2O \\ \qquad\qquad\quad |\quad\ | \\ \qquad\qquad\quad H\ \ CH_2-NH_2 \end{array}$$

One end has $-NH_2$ which can react with the $-COOH$ end of another of these dipeptides.

128.

a.

$$\begin{array}{l} \qquad\qquad\qquad\quad CH_3\ CH_3 \\ \qquad\qquad\qquad\qquad |\qquad\ | \\ CH_3-CH_2-CH-CH-CH-C=O \\ \qquad\qquad\quad\ \ |\qquad\qquad\qquad\ | \\ \qquad\qquad\quad\ Cl\qquad\qquad\qquad H \end{array}$$

b.

$$\begin{array}{l} \qquad\quad NH_2 \\ \qquad\quad\ | \\ CH_3-C-CH_3 \\ \qquad\quad\ | \\ \qquad\quad CH_2Cl \end{array}$$

c.

$$\text{2,4,6-trinitrotoluene: benzene ring with } CH_3 \text{ at C1, } O_2N \text{ at C2, } NO_2 \text{ at C6, } NO_2 \text{ at C4}$$

d.

$$(CH_3)_3C{-}C_6H_4{-}C(CH_3)_3 \quad \text{(para-substituted benzene ring)}$$

e.

$$CH_3{-}CH_2{-}CH_2{-}C(=O){-}O{-}CH_2{-}CH_3$$

f.

$$CH_3{-}CH(Cl){-}CH(CH_3){-}CH(Cl){-}CH(CH_3){-}CH(Cl){-}CH(CH_3){-}CH(Cl){-}CH_3$$

# Chapter Twenty-One Biochemistry

## 21.1 Proteins

2. Molar masses of proteins range from a few thousand amu to over 1 million amu. Such molar masses are consistent with proteins being large polymeric molecules.

4. Globular proteins have an approximately spherical shape. Globular proteins perform functions including the transport and storage of oxygen, the catalysis of many cellular processes, the fighting of infection due to foreign bodies, and the transport of electrons in metabolic processes.

## 21.2 Primary Structure of Proteins

6. The structures of the amino acids are given in detail in Figure 21.2. Generally a side chain is nonpolar if it is mostly hydrocarbon in nature (e.g., alanine, in which the side chain is a methyl group). Side chains are polar if they contain the hydroxyl group (–OH), the sulfhydryl group (–SH), or a second amino ($-NH_2$) or carboxyl (–COOH) group.

8. The primary structure of a protein refers to the specific identity and ordering of amino acids in a protein's polypeptide chain. The primary structure is sometimes referred to as the protein's amino acid *sequence*.

10. 16 (assuming no amino acid repeats)

12. phe–ala–gly

$CH_2$ O $CH_3$ O H

$H_2N$—C—C—N—C—C—N—C—COOH

N terminal H H H H H C terminal

phe–gly–ala

$CH_2$ O H O $CH_3$

$H_2N$—C—C—N—C—C—N—C—COOH

N terminal H H H H H C terminal

ala–phe–gly

$CH_3$ O $CH_2$ O H

$H_2N$—C—C—N—C—C—N—C—COOH

N terminal H H H H H C terminal

ala–gly–phe

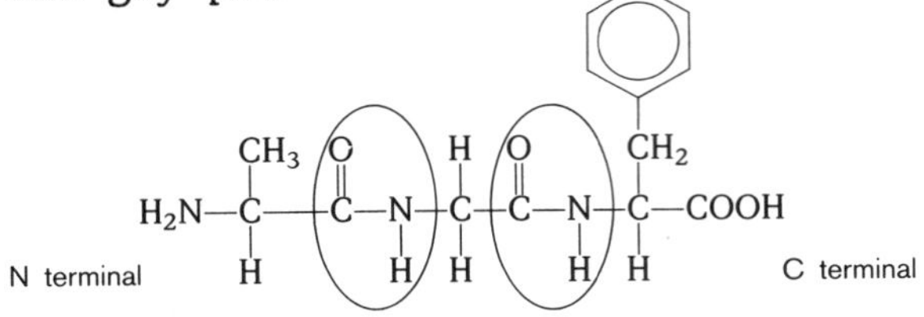

$CH_3$ O H O $CH_2$ (phenyl)

N terminal $H_2N-C-C(=O)-N-C-C(=O)-N-C-COOH$ C terminal

H H H H H

gly–phe–ala

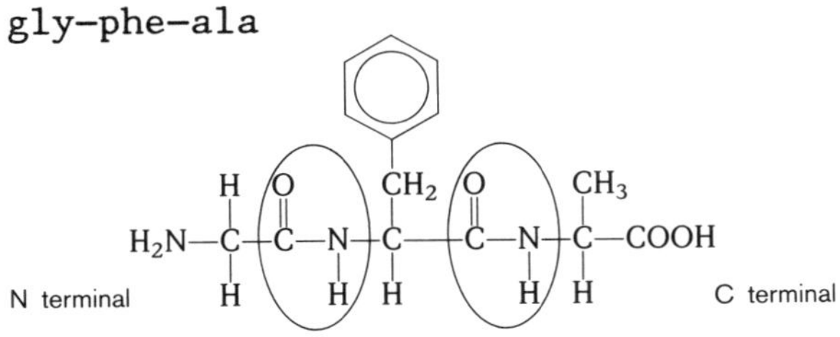

H O $CH_2$ (phenyl) O $CH_3$

N terminal $H_2N-C-C(=O)-N-C-C(=O)-N-C-COOH$ C terminal

H H H H H

gly–ala–phe

H O H O $CH_2$ (phenyl)

N terminal $H_2N-C-C(=O)-N-C-C(=O)-N-C-COOH$ C terminal

H H H H H

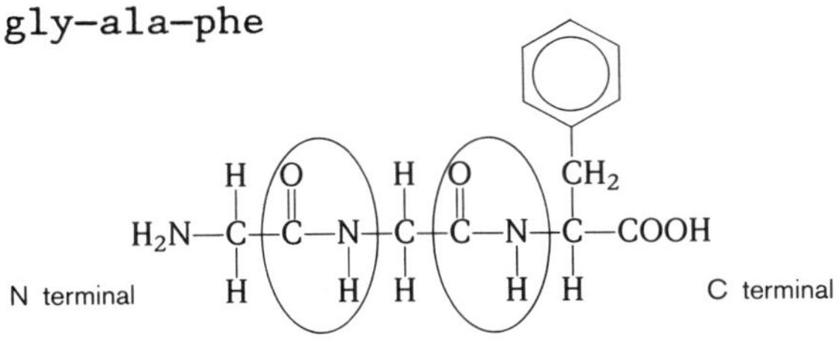

## 21.3 Secondary Structure of Proteins

14. The secondary structure, in general, describes the arrangement of the long polypeptide chain of the protein. In the alpha-helical secondary structure, the chain forms a coil or spiral, which gives proteins consisting of such structures an elasticity or resilience. Such proteins are found in wool, hair, and tendons.

16. Long, thin, resilient proteins, such as hair, typically contain elongated, elastic alpha-helical protein molecules. Other proteins, such as silk, which in bulk form sheets or plates, typically contain protein molecules having the beta pleated sheet secondary structure. Proteins which do not have a structural function in the body, such as hemoglobin, typically have a globular structure.

## 21.4 Tertiary Structure of Proteins

18. Cysteine, an amino acid containing the sulfhydryl (–SH) group in its side chain, is capable of forming disulfide linkages (–S–S–) with other cysteine molecules in the same polypeptide chain. If such a disulfide linkage is formed, this effectively ties together two portions of the polypeptide, producing a kink or knot in the chain, which leads in part to the protein's overall 3-dimensional shape (tertiary structure).

Cysteine, and the disulfide linkages it forms, is responsible for the curling of hair (whether naturally or by a permanent wave).

## 21.5 Functions of Proteins

20. tendons, bone (with mineral constituents), skin, cartilage, hair, fingernails.

22. hemoglobin

24. Proteins contain both acidic (–COOH) and basic ($-NH_2$) groups in their side chains, which can neutralize both acids and bases.

## 21.6 Enzymes

26. substrate

28. An enzyme is inhibited if some other molecule, other than the enzyme's correct substrate, blocks the active sites of the enzyme. If the enzyme is inhibited irreversibly, the enzyme can no longer function and is said to have been inactivated.

## 21.7 Carbohydrates

30. pentoses (5 carbons); hexoses (6 carbons); trioses (3 carbons)

32. A disaccharide consists of two monosaccharide units bound together into a single molecule. Sucrose consists of a glucose molecule and a fructose molecule connected by an alpha-glycosidic linkage.

34. Although starch and cellulose are both polymers of glucose, the glucose rings in cellulose are connected in such a manner that the enzyme which ordinarily causes digestion of polysaccharides is not able to fit the shape of the substrate, and is not able to act upon it.

36. ribose (aldopentose); arabinose (aldopentose); ribulose (ketopentose); glucose (aldohexose); mannose (aldohexose); galactose (aldohexose); fructose (ketohexose).

## 21.8 Nucleic Acids

38. A nucleotide consists of three components: a five-carbon sugar, a nitrogen-containing organic base, and a phosphate group. The phosphate and nitrogen base are each bonded to respective sites on the sugar, but are not bonded to each other.

40. Uracil (RNA only); cytosine (DNA, RNA); thymine (DNA only); adenine (DNA, RNA); guanine (DNA, RNA)

42. When the two strands of a DNA molecule are compared, it is found that a given base in one strand is always found paired with a particular base in the other strand. Because of the shapes and side atoms along the

rings of the nitrogen bases, only certain pairs are able to approach and hydrogen-bond with each other in the double helix. Adenine is always found paired with thymine; cytosine is always found paired with guanine. When a DNA helix unwinds for replication during cell division, only the appropriate complementary bases are able to approach and bond to the nitrogen bases of each strand. For example, for a guanine-cytosine pair in the original DNA, when the two strands separate, only a new cytosine molecule can approach and bond to the original guanine, and only a new guanine molecule can approach and bond to the original cytosine.

44. Messenger RNA molecules are synthesized to be complementary to a portion (gene) of the DNA molecule in the cell, and serve as the template or pattern upon which a protein will be constructed (a particular group of nitrogen bases on *m*-RNA is able to accommodate and specify a particular amino acid in a particular location in the protein). Transfer RNA molecules are much smaller than *m*-RNA, and their structure accommodates only a single specific amino acid molecule: transfer RNA molecules "find" their specific amino acid in the cellular fluids, and bring it to *m*-RNA where it is added to the protein molecule being synthesized.

## 21.9 Lipids

46. A triglyceride typically consists of a glycerine backbone, to which three separate fatty acid molecules are attached by ester linkages

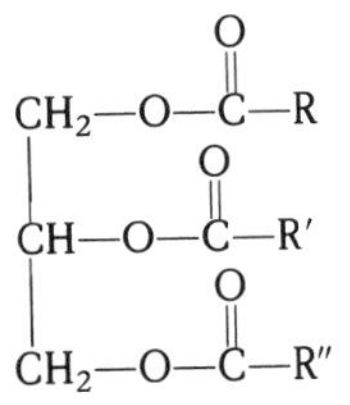

$$\begin{array}{l} CH_2-O-\overset{O}{\overset{\|}{C}}-R \\ \quad | \\ CH-O-\overset{O}{\overset{\|}{C}}-R' \\ \quad | \\ CH_2-O-\overset{O}{\overset{\|}{C}}-R'' \end{array}$$

48. Saponification is the production of a *soap* by treatment of a triglyceride with a strong base such as NaOH.

triglyceride + 3NaOH → glycerol + $3Na^+soap^-$

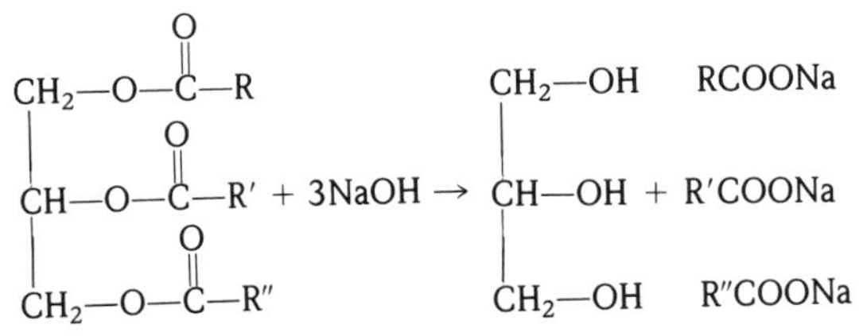

$$\begin{array}{l} CH_2-O-\overset{O}{\overset{\|}{C}}-R \\ \quad | \\ CH-O-\overset{O}{\overset{\|}{C}}-R' \\ \quad | \\ CH_2-O-\overset{O}{\overset{\|}{C}}-R'' \end{array} + 3NaOH \rightarrow \begin{array}{l} CH_2-OH \\ \quad | \\ CH-OH \\ \quad | \\ CH_2-OH \end{array} + \begin{array}{l} RCOONa \\ \\ R'COONa \\ \\ R''COONa \end{array}$$

50. A wax is an ester of a fatty acid with a long chain monohydroxy alcohol. Waxes are solids that provide waterproof coatings on the fruits and leaves of plants, and on the skins and feathers of animals.

52. Cholesterol is the naturally occuring steroid from which the body synthesizes other needed steroids. Since cholesterol is insoluble in water, it is thought that having too large a concentration of this

substance in the bloodstream may lead to its deposition and buildup on the walls of blood vessels, causing their eventual blockage.

54. The primary human bile acid is cholic acid, which helps to emulsify fats in the intestinal tract. In order to be digested by enzymes for absorption into the bloodstream, large clumps of fat must be dispersed into fine droplets in the liquid of the small intestine. Cholic acid basically acts as a detergent.

## Additional Problems

56. i

58. m

60. u

62. f

64. g

66. r

68. p

70. o

72. b

74. d

76. a

78. nucleotides

80. ester

82. adenine, cytosine

84. transfer, messenger

86. insoluble, soluble

88. unsaturated, saturated

90. ionic, nonpolar

92. waxes

94. progesterone